Charles Davies

Differential and Integral Calculus on the Basis of Continuous

Quantity and Consecutive Differences

Charles Davies

Differential and Integral Calculus on the Basis of Continuous Quantity and Consecutive Differences

ISBN/EAN: 9783337811846

Printed in Europe, USA, Canada, Australia, Japan

Cover: Foto ©berggeist007 / pixelio.de

More available books at **www.hansebooks.com**

DIFFERENTIAL

AND

INTEGRAL CALCULUS

ON THE BASIS OF CONTINUOUS QUANTITY AND

CONSECUTIVE DIFFERENCES,

DESIGNED FOR

ELEMENTARY INSTRUCTION,

BY

CHARLES DAVIES, LL.D.,

EMERITUS PROFESSOR OF HIGHER MATHEMATICS IN COLUMBIA COLLEGE.

———

A. S. BARNES & CO., PUBLISHERS,

NEW YORK AND CHICAGO. ◁

1873.

DAVIES' MATHEMATICS.

IN THREE PARTS.

I.—COMMON SCHOOL COURSE.

Davies' Primary Arithmetic.—The fundamental principles displayed in Object Lessons.

Davies' Intellectual Arithmetic.—Referring all operations to the unit 1 as the only tangible basis for logical development.

Davies' Elements of Written Arithmetic.—A practical introduction to the whole subject. Theory subordinated to Practice.

Davies' Practical Arithmetic.—The combination of Theory and Practice, intended to be clear, exact, brief, and comprehensive.

II.—ACADEMIC COURSE.

Davies' University Arithmetic.—Treating the subject exhaustively as *a science*, in a logical series of connected propositions.

Davies' Elementary Algebra.—A connecting link, conducting the pupil easily from arithmetical processes to abstract analysis.

Davies' University Algebra.—For institutions desiring a more complete but not the fullest course in pure Algebra.

Davies' Practical Mathematics.—The science practically applied to the useful arts, as Drawing, Architecture, Surveying, Mechanics, etc.

Davies' Elementary Geometry.—The important principles in simple form, but with all the exactness of rigorous reasoning.

Davies' Elements of Surveying.—Re-written in 1870. A simple and practical presentation of the subject for the scholar and surveyor.

III.—COLLEGIATE COURSE.

Davies' Bourdon's Algebra.—Embracing Sturm's Theorem, and a most exhaustive and scholarly course.

Davies' University Algebra.—A shorter course than Bourdon, for Institutions having less time to give the subject.

Davies' Legendre's Geometry.—The original is the best Geometry of Europe. The revised edition is well known.

Davies' Analytical Geometry.—Being a full course, embracing the equation of surfaces of the second degree.

Davies' Differential and Integral Calculus.—Constructed on the basis of Continuous Quantity and Consecutive Differences.

Davies' Analytical Geometry and Calculus.—The shorter treatises, combined in one volume, as more available for American courses of study.

Davies' Descriptive Geometry.—With application to Spherical Trigonometry, Spherical Projections, and Warped Surfaces.

Davies' Shades, Shadows, and Perspective.—A succinct exposition of the mathematical principles involved.

Davies & Peck's Mathematical Dictionary.—Embracing the definitions of all the terms, and also a Cyclopedia of Mathematics.

Davies' Nature and Utility of Mathematics.—Embracing a condensed Logical Analysis of the entire Science, and of its General Uses.

Entered according to Act of Congress, in the year Eighteen Hundred and Seventy-three, by
CHARLES DAVIES,
In the Office of the Librarian of Congress, at Washington.

PREFACE.

THE Differential and Integral Calculus is too important a branch of Mathematics to be omitted in a course of collegiate instruction.

In the elementary branches, the abstract quantities, Number and Space, are presented to the mind as of definite extent, and as made up of parts.

The value, or measure, in any given case, is expressed by the number of times which the quantity contains one of its parts, regarded as a standard, or unit of measure. But we do not attain to a full and clear apprehension of their *quantitative nature*, until we subject them to the law of continuity, and trace their changes, under this law, as they pass from one state of value to another

The Differential and Integral Calculus embraces all the processes necessary to such an analysis. It regards quantity as the result of change. It examines established laws of change, and determines their consequences. It supposes laws of change, and traces the results of the hypothesis. In short, it embraces within its grasp—in the Material, everything from the minutest atom to the largest body—in Space, all that can be measured, from the geometrical point to absolute infinity—in Time, the entire range of duration—and in Motion, every change from absolute rest to infinite velocity.

The substance of the present volume was published in the year eighteen hundred and sixty. For the want of a proper introduction, its marked characteristics seem not to have attracted public attention. That introduction is now supplied.

The entire system is based on four principles:

1st. Continuous quantity, which is defined;

2d. Consecutive values of continuous quantity, which are defined with reference to the Calculus;

3d. The Differential of a quantity, arising from the two first definitions; viz., *the difference between any two consecutive values of a continuous quantity ;*

4th. That the differential of the independent variable is the common unit of measure for all differentials.

The law of continuity, which may be applied, by hypothesis, to all quantity, not restricted by definition, and which is certainly applicable to time and space, and so far as we know, to all growth and development, is a necessary condition in all operations of the Calculus. Hence, the true theory of the Calculus must be based upon it.

In regard to the present Treatise, it aims to make the law of continuity, in quantity, as accessible and familiar as the law of gravitation in matter. If this be accomplished, the mysteries of the Calculus will disappear; and the subject will embrace, like the other branches of Mathematics, only questions of ratio and measurement. And why should this not be so, after we apprehend, distinctly, the unit of measure and the law of change?

FISHKILL-ON-HUDSON, *January*, 1873.

CONTENTS.

INTRODUCTION.

SECTION I.

DEFINITIONS AND FIRST PRINCIPLES

SECTION II.

DIFFERENTIALS OF ALGEBRAIC FUNCTIONS.

14

SECTION III.

INTEGRATION AND APPLICATIONS.

SECTION IV.

SUCCESSIVE DIFFERENTIALS—SIGNS OF DIFFERENTIAL CO-EFFICIENTS—FORMULAS OF DEVELOPMENT.

SECTION V.

MAXIMA AND MINIMA.

SECTION VI.

DIFFERENTIALS OF TRANSCENDENTAL FUNCTIONS.

SECTION VII.

TRANSCENDENTAL CURVES—CURVATURE—RADIUS OF CURVA-
TURE—INVOLUTES AND EVOLUTES.

INTEGRAL CALCULUS.

INTRODUCTION

DIFFERENTIAL CALCULUS.

1. THE entire science of mathematics is conversant about the properties, relations, and measurement of quantity. Quantity has already been defined. It embraces everything which can be increased, diminished, and measured.

In the elementary branches of mathematics, quantity is regarded as made up of parts. If the parts are equal, each is called a unit, and the measure of a quantity is the number of times which it contains its unit. Such quantities are called *discontinuous ;* because, in passing from one state of value to another, we go by the steps of the unit, and hence, pass over all values lying between adjacent units.

Thus, if we increase a line from one foot to forty feet, by the continued addition of one foot, we touch the line, in our computation, only at its two extremities, and at thirty-nine intermediate points, of which any two adjacent points are one foot apart. In the scale of ascending numbers, 1, 2, 3, 4, 5, 6, etc., we pass over all quantities less than that which is denoted by the unit, one. Discontinuous quantities are generally expressed by numbers, or by letters, which stand for numbers.

2. In the higher branches of mathematics, the laws which regulate and determine the changes of quantity, from one state of value to another, are quite different. Suppose, for example, that instead of considering a right line to be made up of forty feet, or of 480 inches, or of 960 half inches, or of 1,920 quarter inches, or of any number of equal parts of the inch, we regard it as a quantity having its origin at 0, and increasing according to such a law, as to pass through or assume, in succession, all values between 0 and forty feet. This supposition gives us the same distance as before, but a very different law of formation. A quantity so formed or generated, is called a *continuous quantity.* Hence,

A DISCONTINUOUS QUANTITY is one which is made up of parts, and in which the changes, in passing from one state of value to another, can be expressed in numbers, either exactly, or approximatively; and

A CONTINUOUS QUANTITY is one which in changing from one state of value to another, according to a fixed law, passes through or assumes, in succession, all the intermediate values.

Thus, the time which elapses between 12 and 1 o'clock, or between any two given periods, is continuous. All space is continuous, and *every quantity* may be regarded as continuous, which can be subjected to the required law of change.

Limits.

3. THE LIMIT of a variable quantity is a quantity towards which it may be made to approach nearer than any given quantity, and which it reaches, under a particular supposition.

Limits of Discontinuous Quantity.

4. THE limits of a discontinuous quantity are merely numerical boundaries, beyond which the quantity cannot pass.

For positive quantities, the minimum limit is 0, and the maximum limit, infinity. For negative quantities, they are 0, and minus infinity; and generally, using the algebraic language, the limits of all quantities are,

Minimum limit, — infinity; maximum limit, + infinity.

We can illustrate these limits, and also what we mean by the terms, 0 and infinity, plus or minus, by reference to the trigonometrical functions. Thus, when the arc is 0, the sine is 0. When the arc increases to 90°, the sine attains its maximum value, the radius, R. Passing into the second quadrant, the sine diminishes as the arc increases, and when the arc reaches 180°, the sine becomes 0. From that point, to 270°, the sine increases numerically, but *decreases algebraically*, and at 270°, its minimum value is, — R. From 270° to 360°, the sine decreases numerically, but increases algebraically. Hence, the numerical limits of the sine, are 0 and R; and its algebraic limits, — R and + R.

Let us now consider the tangent. For the arc 0, the tangent is 0. If the arc be increased from 0 towards 90°, the length of the tangent will increase, and as the arc approaches 90°, the prolonged radius or secant becomes more nearly parallel with the tangent; and finally, at 90° it becomes absolutely parallel to it, and the length of the

tangent becomes greater than any *assignable* line. Then
we say, that the tangent of 90° is *infinite ;* and we designate that quantity by ∞. After 90°, the tangent becomes
minus, and continues so to the end of the second quadrant,
where it becomes − 0 ; and at 270° it becomes equal to, + ∞.
The secant of 90° is also equal to + ∞ ; and of 270°, to − ∞.
These illustrations indicate the significations of the terms,
zero and *infinity*. They denote the *limits* toward which variable quantities may be made to approach nearer than any
given quantity, and which limits are reached under particular suppositions.

5. The term, *given*, or *assignable quantity*, denotes any
quantity of a limited and fixed value.

The term, *infinitely great*, or *infinity*, denotes a quantity
greater than any assignable quantity of the same kind.

The term, *infinitely small*, or *infinitesimal*, denotes a
quantity less than any assignable quantity of the same
kind.

Continuous Quantities.

6. A continuous quantity has already been defined
(Art. 2.)   By its definition it has two attributes :

1st. That it shall change its value according to a fixed
law ; and

2d. That in changing its value, between any two limits,
it shall pass through all the intermediate values.

7. CONSECUTIVE VALUES.—Two values of a continuous
quantity are consecutive when, if the greater be diminished,

or the less increased, *according to the law of change*, the two values will become equal.

Let A be the origin of a system of rectangular co-ordinate axes, and C a given point on the axis of X.

If we suppose a point to move from A, in the plane of the axes, and with the further condition, that it shall continue at the same distance from the point C, it will generate the circumference of a circle, $APBDEA$, beginning and terminating at the point A. The moving point is called the *generatrix*.

The circumference of this circle may also be generated in another way, thus:

Denote the straight line AD by $2R$, and suppose a point to move uniformly from A to D. Denote the distance from A to any point of the line AD, by x: then, the other segment will be denoted by $2R - x$. Now, *at every point* of AD, suppose a perpendicular to be drawn to AD. Denote each perpendicular by y, and suppose y always to have such a value as to satisfy the equation

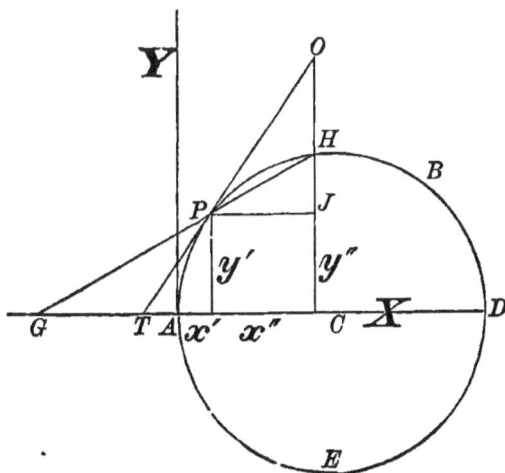

$$y^2 = 2Rx - x^2.$$

Under these hypotheses, it is plain that the extremities of the ordinate y will be found in the circumference of the circle, which will be a continuous quantity. The ordinate y will be contained, in the first quadrant, between the numerical limits of $y = 0$ and $y = + R$; in the second, between the numerical limits of $y = + R$, and $y = 0$; in the third, between $y = - 0$ and $y = - R$; and in the fourth, between $y = - R$ and $y = - 0$.

The circumference $ABDEA$, may be regarded under two points of view:

First. As a discontinuous quantity, expressed in numbers: viz., by $AD \times 3.1416$; or it may be expressed in degrees, minutes, or seconds, viz., $360°$, or $21600'$, or $1296000''$. In the first case, the step, or change, in passing from one value to the next, will be the unit of the diameter AD. In the second, it will be one degree, one minute, or one second. In neither case, will the parts of the circumference less than the unit be reached by the computation. Or,

Secondly: We may regard the circumference as a continuous quantity, beginning and terminating at A. Under this supposition, the generatrix will occupy, in succession, every point of the circumference, and will, in every position, satisfy the equation

$$y^2 = 2Rx - x^2.$$

Hence, if we measure a quantity by a *finite* unit, that quantity is discontinuous: but if we measure it by an *infinitesimal* unit, the quantity becomes continuous.

Tangent Line and Limit.

8. Take any point of the circumference of this circle, as P, whose co-ordinates are x' and y', and a second point H, whose co-ordinates are x'' and y'', and through these points draw the secant line, HPG.

Then, $HJ = y'' - y'$, and $PJ = x'' - x'$; and

$$\frac{HJ}{PJ} = \frac{y'' - y'}{x'' - x'} = \text{tang. of the angle } HPJ, \text{ or } HGC.$$

Let us now suppose a tangent line TP to be drawn to the circle, touching it at P. If we suppose the point H to approach the point P, it is plain that the value of y'' will approach to the value of y', and the value of x'' to that of x': and when the point H becomes consecutive with the point P, y'' and y' will become consecutive, and so also will x'' and x'. When the point H becomes consecutive with the point P, the secant line, HG, becomes the tangent line PT. For, since the arc is a continuous quantity, no point of it can lie between two of its consecutive values; and hence, at P, no point of the curve can lie above the line TP; therefore, by the definitions of Geometry, TP is a tangent line to the circle at the point P.

But the defini- tion of a tangent line to a circle, in elementary Geometry, viz., that it touches the circumfer- ence in one point is incomplete. It is provisional on- ly. For, as we now see, the tan-

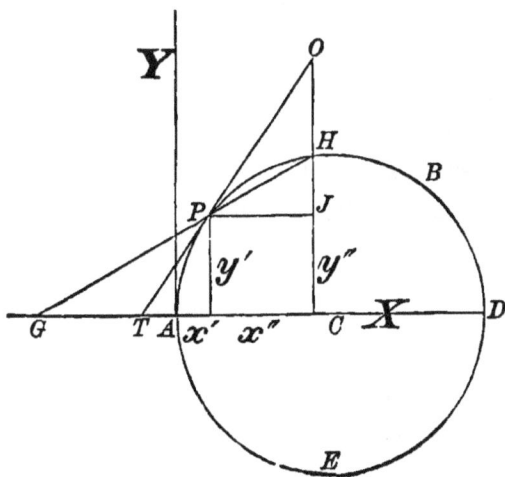

gent line touches the circle in *two consecutive points*, which, in discontinuous quantity, are regarded as one, because the distance between them, expressed numerically, is zero.

If we prolong JH till it meets the tangent line at O, we see that,

$$\frac{JO}{x'' - x'} = \text{tangent of } OPJ = \text{tangent of } OTC; \text{ and that,}$$

$$\frac{JH}{x'' - x'} = \text{tangent of } HPJ = \text{tangent of } HGC.$$

When the point H approaches the point P nearer than any given distance, the angle HGC will approach the angle PTC nearer than any given angle, and when H becomes consecutive with P, the angle HGC will become equal to the angle PTC, which is therefore its *limit*. Under this hypothesis, the point H falls on the tangent line, and JH becomes equal to JO. Under the same hypothesis, y'' and y' become consecutive, and also x'' and x'; hence, $y'' - y'$ becomes less than any given quantity; and so, also, does $x'' - x'$. This difference between consecutive values is expressed by simply writing the letter d before the variable. Thus, the difference of the consecutive values of y, is denoted by dy; and is read *differential* of y; and the difference between the consecutive values of x, is denoted by dx, and is read *differential* of x. Hence, we have

$$\frac{dy}{dx} = \text{tangent } PTC; \text{ viz.,}$$

the tangent of the angle which the tangent at the point P makes the axis of X.

By the definition of a limit, dy becomes the limit of $y'' - y'$, and dx the limit of $x'' - x'$, under the supposition, that y'' and y', and x'' and x' become, respectively, consecutive. The term *limit*, therefore, used to designate the ultimate difference between two values of a variable, denotes the actual difference between its two consecutive values; this difference is infinitely small, and consecutive with zero. For, if after y'' has become consecutive with y', it be again diminished, according to the law of change expressed by the equation

$$y^2 = 2Rx - x^2$$

it will, from the definition of consecutive values, become equal to y', and then x'' will become equal x', and we shall have

$$y'' - y' = 0 \qquad \text{and} \qquad x'' - x' = 0.$$

Under this hypothesis the line PT has, at P, but a *single point*, common with the circumference of the circle; it then ceases to be a tangent, and becomes any secant line passing through this point and intersecting the circumference in a second point.

9. What we have here shown in regard to the circum-

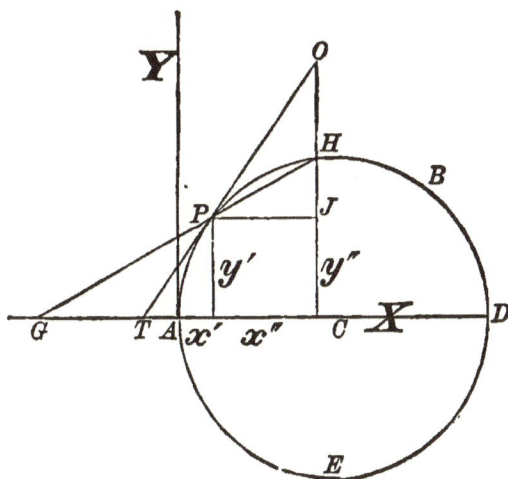

ference of the circle, and its tangent, is equally true of any other curve and its tangent, as may be shown by a very slight modification of the process.

The fact, that a straight line tangent to a curve, has two consecutive points common with it, appears in all the elementary problems of tangents. These conditions, are, an equality between the co-ordinates of the point of contact and the first differential co-efficients, at the same point, of the straight line and curve. These conditions fix the consecutive points common to the straight line and curve.

Analysis, therefore, by its searching and microscopic powers—by looking into the changes which take place in quantity, as it passes from one state of value to another, develops properties and laws which lie beyond the reach of the numerical language. Thus, the distance between two consecutive points, on the circumference of a circle, cannot be expressed by numbers; for, however small the number might be, chosen to express such a distance, it could be diminished, and hence, there would be intermediate points.

The introduction, therefore, of continuous quantity, into the science of mathematics, brought with it new ideas and the necessity of a new language. Quantity, made up of parts, and expressed by numbers, is a very different thing from the continuous quantity treated of in the Differential and Integral Calculus. Here, the law of continuity, in the change from one state of value to another, is the governing principle, and carries with it many consequences.

Time and space are the continuous quantities with which we are most conversant. If we take a moment in time, and look back to the past, or forward to the future, there is no interruption. The law of continuity is unbroken, and the infinite opens to our contemplation. If we take a point in space, and through it conceive a straight line to be drawn, the law of continuity is also there, and the imagination runs along it, to the infinite, in either direction. The attraction of gravitation is a continuous force; and all the motions to which it gives rise, follow the law of continuity. All growth and development, in the vegetable and animal kingdoms, so far as we know, conform to this law. This, therefore, is the great and important law of quantity, and the Higher Calculus is conversant mainly about its development and consequences.

Consequences of the Law of Continuity.

1. The most striking consequence of the law of continuity, is the fact, that whatever be the quantity subjected to this law, or whatever be the law of change, the difference between any two of the consecutive values is an infinitesimal, and hence cannot be expressed by numbers.

2. Since a continuous quantity may be of any value, and be subjected to any law of change, the infinitesimal which expresses the difference between any two of its consecutive values, is a *variable* quantity; and hence, may have any value between zero and its maximum limit.

3. The law of continuity in quantity, therefore, introduces into the science of mathematics a class of variables called *infinitesimals*, or *differentials*. Every variable quantity has, at every state of its value, an infinitesimal corresponding to it. This infinitesimal is the connecting link, in the law of continuity, and will vary with the value of the quantity and the law of change.

4. In the Infinitesimal Calculus, the properties, relations, and measurement of quantities are developed by considering the laws of change to which they are subjected. The elements of the language employed, are symbols of those infinitesimals.

NEWTON'S METHOD OF TREATING CONTINUOUS QUANTITY.*

LEMMA I.

10. *Quantities, and the ratios of quantities, which in any finite time converge continually to equality, and before the end of that time approach nearer, the one to the other, than by any given difference, become ultimately equal.*

If you deny it, suppose them to be ultimately unequal, and let D be their ultimate difference. Therefore, they cannot approach nearer to equality than by that given difference D; which is against the supposition.

LEMMA II.

*If in any figure, AacE, terminated by right lines Aa, AE, and the curve acE, there be inscribed any number of parallelograms Ab, Bc, Cd, etc., comprehended under the equal bases, AB, BC, CD, etc., and the sides Bb, Cc, Dd,
etc., parallel to one side Aa of the figure; and the parallelograms aKbl, bLcm, cMdn, dDEo are completed; then, if the breadth of these parallelograms be supposed to be diminished, and their number to be augmented in finitum; I say, that the ultimate ratios which the inscribed figure AKbLcMdD, the*

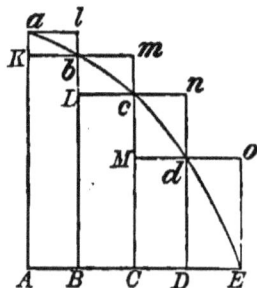

---

* *Principia*, Book I., Section I.

circumscribed figure AalbmcndoE and the curvilinear figure AabcdE will have to one another, are ratios of equality.

For, the difference of the inscribed and circumscribed figures is the sum of the parallelograms, Kl, Lm, Mn, Do, that is, (from the equality of their bases), the rectangle under one of their bases Kb and the sum of their altitudes Aa; that is, the rectangle $ABla$. But this rectangle, because its breadth AB is supposed diminished *in finitum*, becomes less than any given space. And therefore, (by Lemma I.) the figures inscribed and circumscribed, become ultimately equal one to the other; and much more will the intermediate curvilinear figure be ultimately equal to either.

LEMMA III.

11. *The same ultimate ratios are also ratios of equality, when the breadths AB, BC, DC, etc., of the parallelograms are unequal and are all diminished* in finitum.

For, suppose AF to be the greatest breadth, and complete the parallelogram $FAaf$. This parallelogram will be greater than the difference of the inscribed and circumscribed figures; but because its breadth AF is diminished *in finitum*, it will become less than any given rectangle.

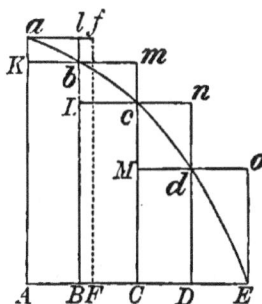

COR. 1. Hence, the ultimate sum of these evanescent parallelograms will, in all parts, coincide with the curvilinear figure.

COR. 2. Much more will the rectilinear figure compre-

hended under the chords of the evanescent arcs, *ab, bc, cd,*
etc. ultimately coincide with the curvilinear figure.

Cor. 3. And also, the circumscribed rectilinear figure
comprehended under the tangents of the same arcs.

Cor. 4. And therefore, these ultimate figures (as to their
perimeters, *acE*) are not rectilinear, but curvilinear limits
of rectilinear figures.

Lemma IV.

12. *If in two figures, AacE, PprT, you inscribe (as before)
two ranks of parallelograms, an equal number in each rank,
and, where their breadths are diminished, in finitum, the
ultimate ratios of the parallelograms in one figure to those
in the other, each to each respectively, are the same; I say,
that those two figures, AacE, PprT, are to one another in
that same ratio.*

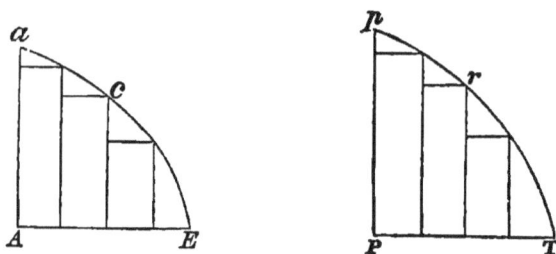

For, as the parallelograms in the one figure are severally
to the parallelograms in the other, so (by composition) is
the sum of all in the one to the sum of all in the other;
and so is the one figure to the other; because (by Lemma
III.), the former figure to the former sum, and the latter
figure to the latter sum, are both in the ratio of equality.

Cor. Hence, if two quantities of any kind are anyhow divided into an equal number of parts, and those parts, when their number is augmented, and their magnitude diminished *in finitum,* have a given ratio one to the other, the first to the first, the second to the second, and so on in order, the whole quantities will be one to the other in that same given ratio. For if in the figures of this lemma, the parallelograms are taken one to the other in the ratio of the parts, the sum of the parts will always be as the sum of the parallelograms; and therefore, supposing the number of the parallelograms and parts to be augmented, and their magnitudes diminished *in finitum,* those sums will be in the ultimate ratio of the parallelogram in the one figure to the corresponding parallelogram in the other; that is (by the supposition), in the ultimate ratio of any one part of the one quantity to the corresponding part of the other.

Lemma V.

13. *In similar figures all sorts of homologous sides, whether curvilinear or rectilinear, are proportional; and their areas are in the duplicate ratio of their homologous sides.*

Lemma VI.

14. *If the arc ACB, given in position, is subtended by the chord AB, and in any point A in the middle of the continued curvature, is touched by a right line AD, produced both ways; then, if the points A and B approach one another and meet, [become consecutive] I say, the angle BAD contained between the chord and the tangent will be diminished* in finitum, *and ultimately will vanish.*

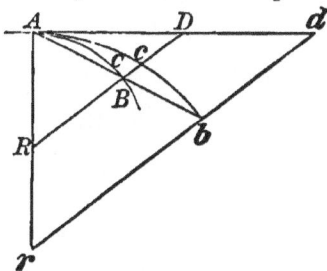

For, if it does not vanish, the arc ACB, will contain with the tangent AD, an angle equal to a rectilinear angle; and therefore, the curvature at the point A will not be continued, which is against the supposition.

LEMMA VII.

15. *The same thing being supposed, I say that the ultimate ratio of the arc, chord, and tangent, any one to any other, is the ratio of equality.*

For, while the point B approaches towards the point A, consider always AB and AD as produced to the remote points b and d, and parallel to the secant BD draw bd: and let the arc Acb be always similar to the arc ACB. Then, supposing the points A and B to coincide, [become consecutive], the angle dAb will vanish, by the preceding lemma; and therefore, the right lines Ab, Ad (which were always finite), and the intermediate arc Acb, will coincide, and become equal among themselves. Wherefore, the right lines AB, AD, and the intermediate arc ACB (which are always proportional to the former), will vanish, and ultimately acquire the ratio of equality.

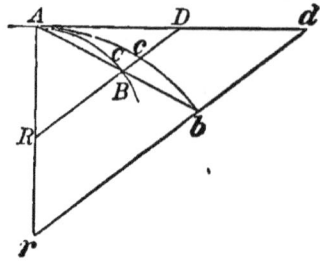

COR. 1. Whence, if through B we draw BF parallel to the tangent, always cutting any right line AF passing through A and F, this line BF will be, ultimately, in the ratio of equality with the evanescent

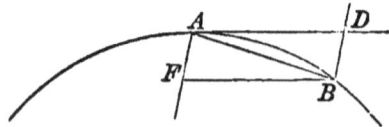

arc ACB; because, completing the parallelogram $AFBD$, it is always in the ratio of equality with AD.

COR. 2. And if through B and A more right lines be drawn BE, BD, AF, AG, cutting the tangent AD and its parallel BF, the ultimate ratio of the abscissas AD, AE, BF, BG, and of the arc AB, any one to any other, will be the ratio of equality.

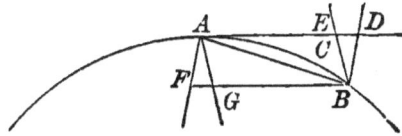

COR. 3. And therefore, in any reasoning about ultimate ratios, we may freely use any one of those lines for any other.

*   *   *   *   *   *

16. *Scholium.*—Those things which have been demonstrated of curve lines, and the superficies which they comprehend, may be easily applied to the curve superficies, and contents of solids. These lemmas are premised to avoid the tediousness of deducing perplexed demonstrations *ad absurdum*, according to the method of the ancient geometers. For demonstrations are more contracted by the method of indivisibles: but because the indivisibles seem somewhat harsh, and therefore, that method is reckoned less geometrical, I chose rather to reduce the demonstrations of the following propositions to the first and last sums and ratios, of nascent and evanescent quantities; that is, to the limits of those sums and ratios; and so, to premise, as short as I could, the demonstration of those limits. For, hereby the same thing is performed as by the method of indivisibles; and now those principles being demonstrated, we may use them with more safety.

Therefore, if hereafter I should happen to consider quantities as made up of particles, or should use little curve lines for right ones, I would not be understood to mean indivisibles, but evanescent divisible quantities; not the sums and ratios of determinate parts, but always the limits of sums and ratios; and that the force of such demonstrations always depends on the method laid down in the foregoing lemmas.

Perhaps it may be objected, that there is no ultimate proportion of evanescent quantities; because the proportion, before the quantities have vanished, is not the ultimate, and when they are vanished, is none. But by the same argument it may be alleged, that a body arriving at a certain place, and there stopping, has no ultimate velocity; because, the velocity, before the body comes to the place, is not its ultimate velocity; when it has arrived, is none. But the answer is easy; for, by the ultimate velocity is meant, that with which the body is moved, neither *before* it arrives at its last place and the motion ceases, nor *after*; but, at the *very instant* it arrives; that is, that velocity with which the body arrives at its last place, and with which the motion ceases. And in like manner, by the ultimate ratio of evanescent quantities is to be understood the ratio of the quantities not before they vanish, not afterwards, but with which they vanish. In like manner, the first ratio of nascent quantities is that with which they begin to be. And the first or last sum, is that with which they begin to be (or to be augmented or diminished). There is a limit which the velocity at the end of the motion may attain, but not exceed. This is the ultimate velocity. And there is the like limit in all quantities and proportions that begin and cease to be. And since such limits are certain and definite, to determine the same is a problem strictly geometrical. But whatever is geometrical we may be allowed to use in determining and demonstrating any other thing that is likewise geometrical.

* * * * * *

Fruits of Newton's Theory.

17. The main difficulties in the higher mathematics, have arisen from inadequate or erroneous notions of ultimate or evanescent quantities, and of the ratios of such quantities. After two hundred years of discussion, of experiment and of trial, opinions yet differ widely in regard to them, and especially in regard to the forms of language by which they are expressed.

One cannot approach this subject, which has so long engaged the earnest attention of the greatest minds known to science, without a feeling of awe and distrust. But tapers sometimes light corners which the rays of the sun do not reach; and as we must adopt a theory in a system of scientific instruction, it is perhaps due to others, that we should assign our reasons therefor.

18. An ultimate, or evanescent quantity, which is the basis of the Newtonian theory, is not the quantity "*before* it vanishes, nor *afterwards;* but, *with which it vanishes.*"

I have sought, in what precedes and follows, to define this quantity—to separate it from all other quantities—to present it to the mind in a crystallized form, and in a language free from all ambiguity; and then to explain how it becomes the key of a sublime science.

As a first step in this process, I have defined continuous quantity (Art. 2.), and this is the only class of quantity to which the Differential Calculus is applicable. The next step was to define consecutive values, and then, the difference between any two of them (Art. 7). These differences are the ultimate or evanescent quantities of Newton. They are not quantities of determinate magnitudes, but such as come from variables that have been diminished indefinitely. They form a class of quantities by themselves, which have their own language and their own laws of change; and they are called, *Infinitesimals*, or *Differentials*.

Since the difference between any two values of a variable quantity, which are near together, but not consecutive, will depend on the relative VALUES of the quantities and the *law* of change, it is plain, that when we pass to the limit of this difference, such limit will also depend for its value on the variable quantity and the law of change: and hence, the infinitesimals are unequal among themselves, and any two of them may have, the one to the other, any ratio whatever.

These infinitesimals will always be quantities of the same *kind* as those from which they were derived; for the *kind* of quantity which expresses a difference, is the same, whether the difference be great or small.

Limits.

19. Marked differences of opinion exist among men of science in regard to the true notion of a *limit;* and hence, definitions have been given of it, differing widely from each other. We have adopted the views of Newton, so clearly set forth in the lemmas and scholium which we have quoted

from the *Principia*. He uses, as stated in the latter part
of the scholium, the term limit, to designate the ultimate
or evanescent value of a variable quantity; and this value is
reached under a particular hypothesis. Hence, our definition (Art. 3).

Let us now refer again to the case of tangency.

Let APB be any curve whatever, and TPF a tangent touching it at the point P. Draw any chord of
the curve, as PB, and though P and B draw the ordinates PD and BH. Also draw PC parallel to TH.

Then, $\dfrac{CF}{PC} =$ tang. $FPC =$ tang. the angle PTH, which
the tangent line TPF makes with the axis TH.

But, $\dfrac{BC}{PC} =$ tangent of the angle
BPC.

If now we suppose BH to move
towards PD, the angle BPC will
approach the angle FPC, which
is its limit. When BH becomes
consecutive with PD, BC will
reach its ultimate value: and since
by Lemma VII., the ultimate
ratio of the arc, chord, and tangent, any one to any other,
is the ratio of equality, it follows that they must then all be
equal, each to each. Under this hypothesis the point B
must fall on the tangent line TPF; that is, the chord and
tangent, in their ultimate state, have two points in common;
hence they coincide; and as the two points of the arc are
consecutive, it must also coincide with the chord and tangent.

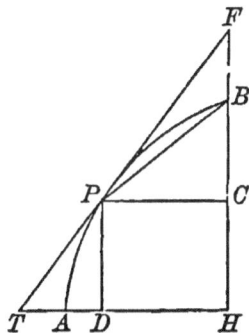

This, at first sight, seems impossible. But if it be granted
that two points of a curve can be consecutive and that a
straight line can be drawn through *any* two points, we have
the solution. If we deny that two points of the curve can be
consecutive, we deny the law of continuity.

The method of Leibnitz adopted the simple hypothesis
that when the point B approached the point P, infinitely
near, the lines CF and CB become infinitely small, and that
then, either may be taken for the other; under which
hypothesis the ratio of PC to CB, becomes the ratio of PC
to CF.

What the Lemmas of Newton prove.

20. The first lemma, which is "the corner-stone and
support of the entire system," predicates *ultimate* equality
between any two quantities which continually approach
each other in value, and under such a law of change, that,
in any finite time they shall approach nearer to each other
than by any given difference. The common quantity towards
which the quantities separately converge, is the limit of each
and both of them, and this limit is always reached under a
particular supposition.

Lemmas II., III., and IV. indicate the steps by which we
pass from discontinuous to continuous quantity. They
introduce us, fully, to the law of continuity. They dem-
onstrate the great truth, that the curvilinear space is the
common limit of the inscribed and circumscribed parallelograms, and that this limit is reached under the hypothesis
that the breadth of each parallelogram is infinitely small,
and the number of them, infinitely great. Thus, we reach

the law of continuity; and each parallelogram becomes a connecting link, in passing from one consecutive value to another, when we regard the curvilinear area as a variable. That there might be no misapprehension in the matter, corollary 1, of Lemma III., affirms, that, " the ultimate sum of these evanescent parallelograms, will, in all parts, coincide with the curvilinear figure." Corollary 4, also, affirms that, "therefore, these ultimate figures (as to their perimeters, acE), are not rectilinear, but curvilinear limits of rectilinear figures:" that is, the curvilinear area AEa is the common limit of the inscribed and circumscribed parallelograms, and the curve $Edcba$, the common limit of their perimeters. This can only take place when the ordinates, like Dd, Cc, Bk, become consecutive; and then, the points o, n, m, and l fall on the curve.

The law of continuity carries with it, necessarily, the ideas of the infinitely small and the infinitely great. These are correlative ideas, and in regard to quantity, one is the reciprocal of the other. The inch of space, as well as the curved line, or the curvilinear surface of geometry, has within it the seminal principles of this law.

If we regard it as a continuous quantity, having increased from one extremity to the other, without missing any point of space, we have, the law of change, the infinitely small (the difference between two consecutive values, or the link in the law of continuity), and the infinitely great, in the number of those values which make up the entire line.

It has been urged against the demonstrations of the lemmas, that a mere *inspection* of the figures proves the demonstrations to be wrong. For, say the objectors, there will be, always, *obviously,* "a portion of the exterior parallelograms lying without the curvilinear space." This is certainly true for any *finite* number of parallelograms.

But the demonstrations are made under the express hypothesis, that, "the breadth of these parallelograms be supposed to be diminished, and their number to be augmented, *in finitum.*" Under this supposition, as we have seen, the points, *o, n, m,* and *l,* fall in the curve, and then the areas named are certainly equal.

Newton's Method in harmony with that of Leibnitz.

21. The method of treating the Infinitesimal Calculus, by Leibnitz, subsequently amplified and developed by the Marquis L'Hopital, is based on two fudamental propositions, or demands, which were assumed as axioms.

I. That if an infinitesimal be added to, or subtracted from, a finite quantity, the sum or difference will be the same as the quantity itself. This demand assumes that the infinitesimal is so small that it cannot be expressed by numbers.

II. That a curved line may be considered as made up of an infinite number of straight lines, each one of which is infinitely small.

It is proved in Lemma II. that the sum of the ultimate rectangles *Ab, Bc, Cd, Do,* etc., will be equal to the curvilinear area *AaE.* This can only be the case when each is "less than any given space," and their number infinite. What is meant by the phrase, "becomes less than any given space?" Certainly, a space too small to be expressed by numbers; for, if we have such a space, so expressed, we can diminish it by diminishing the number, which would be contrary to the hypothesis. This ultimate value, then, of either of the rectangles, is numerically zero: and hence, its addition to, or subtraction from, any finite quantity, would not change the value. The ultimates of Newton, therefore, conform to the first demand of Leibnitz, as indeed they should do; for, they are not numerical quantities, but connecting links in the law of continuity.

It is proved in lemma VII., that the ultimate ratio of the arc, chord, and tangent, any one to any other, is the ratio of equality: hence, their ultimate values are equal. When this takes place, the two extremities of the chord become consecutive, and the remote extremity of the tangent falls on the curve, and coincides with the remote extremity of the chord: that is, F falls on the curve, and PB and PF, coincide with each other, and with the curve. The length of this arc, chord, or tangent, in their ultimate state, is

$$\sqrt{dx^2 + dy^2},$$

a value familiar to the most superficial student of the Calculus.

Behold, then, one side of the inscribed polygon, when such side is infinitely small, and the number of them infinitely great.

That such quantities as we have considered, have a conceivable existence as subjects of thought, and do or may have, *proximatively*, an *actual* existence, is clearly stated in the latter part of the scholium quoted from the *Principia*. It is there affirmed: " This is the ultimate velocity. And there is a *like limit* in all *quantities* and proportions which *begin* and *cease* to be. And since such limits are *certain* and *definite,* to determine the same is a problem *strictly geo-*

metrical. But whatever is geometrical we may be allowed to use in determining and demonstrating any other thing that is likewise geometrical." * Hence, the theory of Newton conforms to the second demand in the theory of Leibnitz.

Different Definitions of a Limit.

22. The common impression that mathematics is an exact science, founded on axioms too obvious to be disputed, and carried forward by a logic too luminous to admit of error, is certainly erroneous in regard to the Infinitesimal Calculus. From its very birth, about two hundred years ago, to the present time, there have been very great differences of opinion among the best informed and acutest minds of each generation, both in regard to its fundamental principles and to the forms of logic to be employed in their development. The conflicting opinions appear, at last, to have arranged themselves into two classes; and these differ, mainly, on this question: What is the correct apprehension and right definition of the word *limit?* All seem to agree that the methods of treating the Calculus must be governed by a right interpretation of this word. The two definitions which involve this conflict of opinion, are these:

1. *The limit of a variable quantity is a quantity towards which it may be made to approach nearer than any given quantity and which it reaches under a particular supposition.*

And the following definition, from a work on the Infinitesimal Calculus by M. Duhamel, a French author of recent date:

---

* NOTE,—The italics are added; they are not in the text.

2. The limit of a variable is *the constant quantity which the variable indefinitely approaches, but never reaches.*

This definition finds its necessary complement in the following definition by the same author:

"We call," says he, "an infinitely *small quantity, or simply, an infinitesimal, every variable magnitude of which the limit is zero.*

The difference between the two definitions is simply this: by the first, the variable, ultimately, reaches its limit; by the second, it approaches the limit, but never reaches it. This apparently slight difference in the definitions, is the dividing line between classes of profound thinkers; and whoever writes a Calculus or attempts to teach the subject, must adopt one or the other of these theories. The first is in harmony with the theories of Leibnitz and Newton, which do not differ from each other in any important particular. It seems also to be in harmony with the great laws of quantity. In discontinous quantity, especially, we certainly include the limits in our thoughts, and in the forms of our language. When we speak of the quadrant of a circle, we include the arc zero and the arc of ninety degrees. Of its functions, the limits of the sine, are zero and radius; zero for the arc zero, and radius for the arc of ninety degrees. For the tangents, the limits are zero and infinity; zero for the arc zero, and infinity for the arc of ninety degrees; and similarly for all the other functions. For all numbers, the limits are zero and infinity; and for all algebraic quantities, minus infinity and plus infinity.

When we consider continuous quantity, we find the second definition in direct conflict with the first Lemma of Newton, which has been well called, "the corner-

stone and foundation of the *Principia.*" It is very
difficult to comprehend that two quantities may approach each other in value, and in any given time become
nearer equal than any given quantity, and yet never become
equal; not even when the approach can be continued to
infinity, and when the law of change imposes no limit to
the decrease of their difference. This, certainly, is contrary
to the theory of Newton.

Take, for example, the tangent line to a curve, at a given
point, and through the point of tangency draw any secant,
intersecting the curve, in a second point. If now, the
second point be made to approach the point of tangency,
both definitions recognize the angle which the tangent line
makes with the axis of abscissas as the limit of the angles
which the secants make with the same axis, as the second
point of secancy approaches the tangent point. By the first
definition, the supposition of consecutive points causes the
secant line to coincide with, and become the tangent. But
by the second definition, the secant line can never become
the tangent, though it may approach to it as near as we
please. This is in contradiction to all the analytical
methods of determining the equations of tangent lines to
curves. See corollaries 1, 2, 3, and 4 of lemma III.; in which
all the quantities referred to are supposed to reach their
limits.

By the second definition, there would seem to be an im-
passable barrier placed between a variable quantity and its
limit. If these two quantities are thus to be forever sepa-
rated, how can they be brought under the dominion of a
common law, and enter together into the same equation.
And if they cannot, how can any property of the one be

used to establish a property of the other? The mere fact of approach, though infinitely near, would not seem to furnish the necessary conditions.

The difficulty of treating the subject in this way is strikingly manifested in the supplementary definition of an infitesimal. It is defined, simply, as " *every variable magnitude whose limit is zero.*"

Now, may not zero be a limit of every variable which has not a special law of change? Is not this definition too general to give a DEFINITE idea of the individual thing defined—an infinitisimal? We have no crystallized notions of a class, till we apprehend, distinctly, the individuals of the class—their marked characteristics—their harmonies and their differences; and also, their laws of relation and connection.

Having given and illustrated these definitions, M. Duhamel explains the methods by which we can pass from the infinitesimals to their limits; and when, and under what circumstances, those limits may be substituted and used for the quantities themselves. Those methods have not seemed to me as clear and practical as those of Newton and Leibnitz.

It is essential to the unity of mathematical science, that all the definitions, should, as far as possible, harmonize with each other. In all discontinuous quantities, the boundaries are included, and are the proper limits. In the hyperbola, for example, we say that the asymptote is the limit of all tangent lines to the curve. But the asymptote is the tangent, when the point of contact is at an infinite distance from the vertex: and any tangent will become the asymptote, under that hypothesis.

If s denotes any portion of a plane surface, y, the ordinate and x the abscissa, we have the known formula:

$$ds = ydx.$$

If we integrate between the limits of $x = 0$, and $x = a$, we have, by the language of the Calculus

$$\int_c^a ds = \int y\, dx,$$

which is read, "integral of the surface between limits of $x = 0$, and $x = a$," in which both boundaries enter into the result.

The area, actually obtained, begins where $x = 0$, and terminates where $x = a$, and not at values *infinitely* near those limits.

What Quantities are denoted by 0.

23. Our acquaintance with the character 0, begins in Arithmetic, where it is used as a necessary element of the arithmetical language, and where it is entirely without value, meaning, absolutely nothing. Used in this sense, the largest finite number multiplied by it, gives a product equal to zero; and the smallest finite number divided by it, gives a quotient of infinity.

When we come to consider variable and continuous quantity, the infinitesimal or element of change from one consecutive value to another, is not the zero of Arithmetic, though it is smaller than any number which can be expressed in terms of one, the base of the arithmetical system.

Hence, the necessity of a new language. If the variable is denoted by x, we express the infinitesimal by dx; if by y, then by dy; and similarly, for other variables.

Now, the expressions dx and dy, have no exact synonyms in the language of numbers. As compared with the unit 1, neither of them can be expressed by the smallest finite part of it. Hence, when it becomes necessary to express such quantities in the language of number, they can be denoted only by 0. Therefore, this 0, besides its first function in Arithmetic, where it is an element of language, and where the value it denotes is absolutely nothing, is used, also, to denote the numerical values of the infinitesimals. Hence, it is correctly defined as a character which sometimes denotes absolutely nothing, and sometimes an infinitely small quantity. We now see, clearly, what appears obscure in Elementary Algebra, that the quotient of zero divided by zero, may be zero, a finite quantity, or infinity.

Inscribed and circumscribed Polygons unite on the Circle.

24. The theory of limits, developed by Newton, is not only the foundation of the higher mathematics, but indicates the methods of using the Infinitesimal Calculus in the elementary branches. This Calculus being unknown to the ancients, their Geometry was encumbered by the tedious methods of the *reductio ad absurdum*. Newton says in the scholium: " These lemmas are premised to avoid the tediousness of deducing perplexed demonstrations *ad absurdum*, according to the method of the ancient geometers."

Lemma I., which is the " corner-stone and foundation of the *Principia*," is also the golden link which connects geometry with the higher mathematics.

It is demonstrated in Euclid's Elements, and also in Davies' *Legendre*, Book V., Proposition X., that " *Two regular
polygons of the same number of sides can be constructed, the
one circumscribed about the circle and the other inscribed
within it, which shall differ from each other by less than any
given surface.*"

The moment it is proved that the exterior and interior
polygons may be made to differ from each other by less
than any given surface, Lemma I. steps in and affirms an ultimate equality between them. And when does that ultimate
equality take place, and when and where do they become
coincident? Newton, in substance affirms, in his lemmas,
"on their common limit, the circle," and under the same
hypothesis as causes the inscribed and circumscribed parallelograms to become equal to their common limits, the
curvilinear area. If Lemma I. is true, the perimeters of the
two polygons will ultimately coincide on the circumference
of the circle, and become equal to it. But what then is the
side of each polygon? We answer, the distance between
two consecutive points of the circumference of the circle?
And what is that value? We answer, the $\sqrt{dx^2 + dy^2}$.

But it is objected, that this introduces us to the infinitely
small. True, it does; but we cannot reach a continuous
quantity without it. The sides of the polygons, *so long as
their number is finite*, will be straight lines, each diminishing
in value as their number is increased. While this is so, the
perimeter of each will be a discontinuous quantity, made up
of the equal sides, each having a finite value, and each being
the unit of change, as we go around the perimeter. As each of
these sides is diminished in value, and their number increased, the discontinuous quantity approaches the law of conti-

nuity, which it reaches, under the hypothesis, that each side becomes infinitely small and their number infinitely great. Behold the polygons embracing each other on their common limit, the circle, and the perimeter of each coinciding with the circumference. Thus, the principles of the Infinitesimal Calculus take their appropriate place in Elementary Geometry, to the exclusion of the cumbrous methods of the *reductio ad absurdum* of the ancients, and the whole science of Mathematics is brought into closer harmonies and nearer relations.

Differential and Integral Calculus.

25. We have seen that the Differential and Integral Calculus is conversant about continuous quantity. We have also seen, that such quantities are developed by considering their laws of change. We have further seen, that these laws of change are traced by means of the differences of consecutive values, taken two and two, as the variables pass from one state of value to another. Indeed, those differences are but the foot-steps of these laws.

Language of the Calculus.

26. We are now to explain the language by which the quantities are represented, by which their changes are indicated, and by which their laws of change are traced. The constant quantities which enter into the Calculus are represented by the first letters of the alphabet, a, b, c, etc., and the variables, by the final letters, x, y, z, etc.

When two variable quantities, y and x, are connected in an equation, either of them may be supposed to increase or decrease *uniformly;* such variable is called the *independent*

2*

variable, because the *law of change is arbitrary*, and *independent* of the form of the equation. This variable is generally denoted by x, and called simply, the *variable*. Under this hypothesis, the change in the variable y will depend on the *form* of the equation: hence, y is called the *dependent* variable, or *function*. When such relations exist between y and x, they are expressed by an equation of the form

$$y = F(x), \qquad y = f(x), \text{ or, } \qquad f(x, y) = 0,$$

which is read, y a function of x. The letter F, or f, is a mere symbol, and stands for the word *function*. If y is a function of x, that is, changes with it, x may, if we please, be regarded as a function of y; hence,

One quantity is a function of another, when the two are so connected that any change of value, in either, produces a corresponding change in the other.

It has been already stated (Art. 8), that the difference between two consecutive values of a variable quantity, is indicated by simply writing the letter d as a symbol, before the letter denoting that variable; so that dx denotes the difference between two consecutive values of the variable quantity denoted by x, and dy the difference between the *corresponding* consecutive values of the variable quantity denoted by y. These are mere forms of language, expressing laws of change.

How are the changes in these variable quantities, expressed by the infinitesimals, to be measured? Only by taking one of them as a standard—and finding how many times it is contained in the other.

The independent variable is always supposed to *increase*

uniformly; hence, the difference between any two of its consecutive values, taken at pleasure, is the same : therefore, this difference, which does not vary in the same equation, or under the same law of change, affords a convenient standard, or unit of measure, and in the Calculus, is always used as such.

The change in the function y, denoted by dy, is always compared with the *corresponding* change of the independent variable, denoted by dx, as a standard, or unit of measure. But the change in any quantity, divided by the unit of measure, gives the *rate* of change: hence, $\dfrac{dy}{dx}$ is the rate of change of the function y. This rate of change is called the *differential coefficient* of y regarded as a function of x, and performs a very important part in the Calculus. The quantities dy and dx, being both infinitesimals, are of the same species : hence, their quotient is an *abstract number*. Therefore, the differential coefficient is a connecting link between the infinitesimals and numbers.

If any quantity whatever be divided by its unit of measure, the quotient will be an abstract number; and if this quotient be multiplied by the unit of measure, the product will be the concrete quantity itself. Hence, if we multiply $\dfrac{dy}{dx}$, by the unit of measure dx, we have $\dfrac{dy}{dx}\,dx$, which always denotes the difference between two consecutive values of y; and therefore, is the differential of y. Hence, *the differential of a variable function is equal to the differential coefficient multiplied by the differential of the independent variable.*

The method, therefore, of dealing with infinitesimals, is

---

precisely the same as that employed for discontinuous quantities.

We assume a unit of measure which is as arbitrary as *one*, in numbers, or, as the foot, yard, or rod, in linear measure, and then we compare all other infinitesimals with this standard. We thus obtain a ratio which is an abstract number, and if this be multiplied by the unit of measure, we go back to the concrete quantity from which the ratio was derived.

We have thus sketched an outline of the Infinitesimal Calculus. We have named the quantities about which it is conversant, the laws which govern their changes of value, and the language by which these laws are expressed. We have found here, as in the other branches of mathematics, that an arbitrary quantity, assumed as a unit of measure, is the base of the entire system; and that the system itself is made up of the various processes employed in finding the ratio of this standard, to the quantities which it measures.

DIFFERENTIAL CALCULUS.

SECTION I.

DEFINITIONS AND FIRST PRINCIPLES.

Definitions.

1. In the Differential Calculus, as well as in Analytical Geometry, the quantities considered are divided into two classes:

1st. *Constant quantities, which preserve the same values in the same investigation;* and,

2d. *Variable quantities, which assume all possible values that will satisfy any equation which expresses the relation between them.*

The constants are denoted by the first letters of the alphabet, a, b, c, &c.; and the variables, by the final letters, x, y, z, &c.

Uniform and varying changes.

2. There are two ways in which a variable quantity may pass from one value to another.

If the variable x, once had the particular value, $x = a$, and afterwards assumed the value, $x = a'$, we can suppose:

1st. That during the change from a to a', x assumed, in succession, and by a uniform change, *all the values* between a and a', just as a body moving uniformly over a given straight line passes through all the points between its extremities; or,

2d. We may suppose, that during the change from a to a', x assumed all possible values between its limits, without the condition of a uniform change. In both cases, the quantity is said to be *continuous*.

3. If two variable quantities, y and x, are connected in an equation, as, for example,

$$y = x^2 + 2;$$

then, to every value of x, arbitrarily assumed, there will be a corresponding value of y, *dependent upon, and resulting from*, the value attributed to x. Thus, if we make $x = 4$, we have,

$$y = 16 + 2 = 18.$$

If we suppose x to increase from 4 to 5, we shall have,

$$y = 25 + 2 = 27;$$

thus, while x changes from 4 to 5, y changes from 18 to 27.

If now we suppose x to increase from 5 to 6, y will increase from 27 to 38. Thus, while x increases uniformly by 1, y will change its value according to a very different law.

Function and variable.

4. When two variable quantities, y and x, are connected in an equation, either of them may be supposed to increase or decrease uniformly; a variable, so changing, is called the *independent variable*, because the *law of change is arbitrary*, and independent of the form of equation. This variable is generally denoted by x, and called simply, the *variable*. The change in the variable y, depends on the *form* of the equation; hence, y is called the *dependent* variable, or *function*. When such a relation exists between y and x, it is expressed by an equation of the form,

$$y = F(x), \quad y = f(x); \quad \text{or,} \quad f(y, x) = 0;$$

which is read, y a function of x. The letter F, or f, is a mere symbol, and stands for the word, *function*. If y is a function of x, that is, changes with it, x is also a function of y; hence,

One quantity is a function of another, when the two are so connected that any change of value, in either, produces a corresponding change in the other.

5. If the equation connecting y and x, is of such a form that y occurs *alone*, in the first member, y is called an *explicit* function of x. Thus, in the equations,

$$y = ax + b \ . \ . \ . \ . \ . \text{ of a straight line,}$$

$$y = \sqrt{R^2 - x^2} \ . \ . \ . \ . \text{ of the circle,}$$

$$y = \frac{B}{A} \sqrt{A^2 - x^2} \ . \ . \ . \text{ of the ellipse,}$$

$$y = \sqrt{2px} \ . \ . \ . \ . \ . \text{ of the parabola, and}$$

$$y = \frac{B}{A} \sqrt{x^2 - A^2} \ . \ . \ . \text{ of the hyperbola,}$$

y is an *explicit* function of x.

But, if the equations are written under the forms,

$$y - ax - b = 0; \qquad \text{or,} \qquad f(x, y) = 0,$$
$$y^2 + x^2 - R^2 = 0; \qquad \text{or,} \qquad f(x, y) = 0,$$
$$A^2 y^2 + B^2 x^2 - A^2 B^2 = 0; \qquad \text{or,} \qquad f(x, y) = 0,$$
$$y^2 - 2px = 0; \qquad \text{or,} \qquad f(x, y) = 0,$$
$$A^2 y^2 - B^2 x^2 + A^2 B^2 = 0; \qquad \text{or,} \qquad f(x, y) = 0,$$

y is called an *implicit* function of x; the nature of the relation between y and x being *implied*, but not developed in the equation.

6. It is plain, that in each of the above equations the *absolute* value of y, for any given value of x, will depend on the constants which enter into the equation; this relation is expressed, by calling y an *arbitrary* function of the constants on which it depends. Thus, in the equation of the straight line, y is an arbitrary function of a and b; in the equation of the circle, y is an arbitrary function of R; in the equation of the ellipse, of A and B; in the equation of the parabola, of $2p$; and in the equation of the hyperbola, of A and B.

7. An *increasing* function is one which increases when the variable increases, and decreases when the variable decreases. A *decreasing* function is one which decreases when the variable increases, and increases when the variable decreases.

In the equation of a straight line, in which a is positive, y is an increasing function of x. In the equations of the circle and ellipse, y is a decreasing function of x. In the equation of the parabola, y is an increasing function of x. In the equation of the hyperbola, y is imaginary for all values of $x < A$, and an increasing function for all positive values of $x > A$.

8. A quantity may be a function of two or more variables. If

$$u = ax + by^2, \quad \text{or} \quad u = ax^2 - by^3 + cz + d,$$

u will be a function of x and y, in the first equation, and of x, y, and z, in the second. These expressions may be thus written:

$$u = f(x, y), \quad \text{and} \quad u = f(x, y, z).$$

If, in the second equation, we make, in succession, the independent variables x, y, and z, respectively equal to 0, we have,

for, $x=0$, $\qquad\qquad\qquad u = -by^3 + cz + d = f(y, z)$,

for, $x=0$, and $y = 0$, $\quad u = cz + d \qquad = f(z)$; and,

for, $x=0$, $y=0$, and $z = 0$, $u = d \qquad\qquad =$ a constant.

Algebraic and Transcendental Functions.

9. There are two general classes of functions: *Algebraic* and *Transcendental*.

Algebraic functions are those in which the relation between the function and the variable can be expressed in the language of Algebra alone: that is, by addition, subtraction, multiplication, division, the formation of powers denoted by constant exponents, and the extraction of roots indicated by constant indices.

Transcendental functions are those in which the relation between the function and variable cannot be expressed in the language of Algebra alone. There are three kinds:

1. *Exponential* functions, in which the variable enters as an exponent; as,

$$u = a^x.$$

2. *Logarithmic* functions, which involve the logarithm of the variable; as,

$$u = \log x.$$

3. *Circular* functions, which involve the arc of a circle, or some function of the arc; as,

$$u = \sin x, \quad u = \cos x, \quad u = \tan x.$$

Geometrical representation of Functions.

10. With the aid of Analytical Geometry, it is easy to trace, geometrically, the numerical relation between any function and its independent variable.

Suppose we have given the equation,

$$y = f(x).$$

If we attribute to x, the independent variable, in succession, every value between $-\infty$ and $+\infty$, each will give a corresponding value for y, which may be determined from the equation, $y = f(x)$

Let O be the origin of a system of rectangular co-ordinates. From O, lay off to the right, all the positive values of x, and to the left all the negative values. Through the extremity of each abscissa, so determined, draw a line parallel to the axis of ordinates, and equal to the corresponding value of y; the plus values

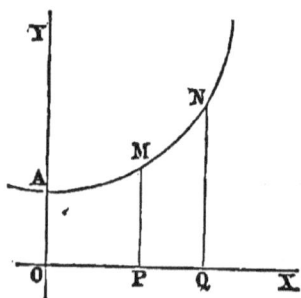

will fall above the axis of X, and the negative values below it; then trace a curve, AMN, through the extremities of these ordinates. The co-ordinates of this curve will indicate every relation between y and x, expressed by the equation,

$$y = f(x).$$

This curve should present to the mind, not merely any particular value of x, and the corresponding value of y, but the entire *series of corresponding values* of these two variables.

Quantities infinitely small—Differentials.

11. A quantity is *infinitely small*, when it cannot be diminished, according to the law of change, without becoming 0.

If, in the equation of the curve,

$$y = f(x) \ . \ . \ (1.)$$

x has a particular value OP, y will denote the ordinate PM.

If x be increased by PQ, de- noted by h, PM will change to QN, which we will denote by y'; and we shall have,

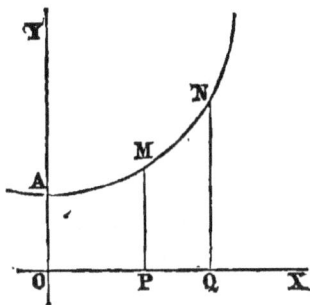

$$y' = f(x + h) \ . \ . \ . \ . \ . \ (2.)$$

If we subtract equation (1) from (2), we obtain,

$$y' - y = f(x + h) - f(x) \ . \ . \ . \ (3.)$$

It is evident that each member of this equation will reduce to 0, when we make $h = 0$.

Let us suppose, as before, that the abscis x has increased from OP to OQ, and that the corresponding ordinate y, has become y'. Draw through the points N and M, the secant line NM. If, now, we suppose the point N to approach M, till it becomes consecutive with it, then,

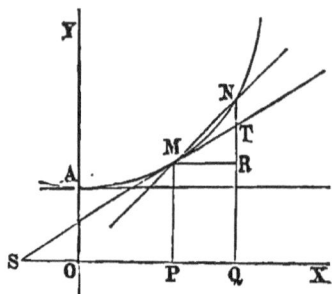

1. The secant line will become the tangent SMT;
2. The abscissas OP and OQ will become consecutive;
3. The ordinates PM, QN, will also be consecutive.

The DIFFERENTIAL of a quantity is the difference between any two of its *consecutive* values; hence, it is indefinitely small.

The differential is expressed, by writing d before the letter denoting the quantity: thus, dx denotes the differential of x, and is read, differential of x: dy denotes the differential of y, and is read, differential of y.

It is plain, that dx denotes the *last* value of h, in Equation (3), before it becomes 0; and that dy denotes the *last* difference between y' and y, as h approaches to 0.

Differential Coefficient.

12. Under the preceding hypotheses, the differentials of x and y admit of geometrical interpretations.

If we divide both members of Equation (3) by h, we have,

$$\frac{y' - y}{h} = \frac{f(x + h) - f(x)}{h} \quad . \quad . \quad . \quad (4)$$

Having drawn MR parallel to the axis of abscissas, NR will denote the difference of the two ordinates y' and y; hence, the first, and consequently the second member of Equation (4), will denote the tangent of the angle NMR, which the secant line makes with the axis of X. Denote the angle which the tangent makes with the axis of X by α. When the ordinates y' and y become consecutive, the secant NM becomes tangent to the curve at the point M, and the angle NMR becomes equal to TSP; and we have,

$$\frac{dy}{dx} = \tan \alpha \quad . \quad . \quad . \quad . \quad (5.)$$

The term $\frac{dy}{dx}$, is called the *differential coefficient* of y, regarded as a function of x; hence,

The differential coefficient of a function is the differential of the function divided by the differential of the independent variable.

If we multiply both members of Equation (5) by dx,

$$\frac{dy}{dx}dx = \tan \alpha \; dx;$$

but the tan α multiplied by the base dx, of the indefinitely small triangle, is equal to the perpendicular, which is the difference between the consecutive values of y' and y, denoted by dy; therefore,

$$\frac{dy}{dx}dx = dy; \text{ hence,}$$

The differential of a function is equal to its differential coefficient multiplied by the differential of the variable.

Limiting Ratio.

13. Let us now resume the consideration of Equation (4).

$$\frac{y' - y}{h} = \frac{f(x + h) - f(x)}{h} \quad . \quad . \quad (4.)$$

The first member of this equation is the ratio of the increment h, of the independent variable, to the correspond ing increment of the function, and denotes the tangent of the angle which the secant line, drawn through the extremities of y and y', makes with the axis of abscissas.

If we suppose h to decrease, the secant line will approach the tangent, and the ratio will approach the tangent of the angle which the tangent line makes with the axis of X. The tan α, is, therefore, the *limit* of this ratio, and since it is also the differential coefficient, it follows that,

The differential coefficient is the limit of the ratio of the increment of the independent variable to the increment of the function.

A varying ratio, of any increment of the independent variable denoted by h, to the corresponding increment of the function, denoted by $y' - y$, reaches its limit when h reaches its last value; and then, the values of y' and y become *consecutive;* therefore, the *limiting* ratio, is *the ratio of consecutive values.* Hence, if we have an expression for the ratio of the increments, we pass to the limiting ratio, or differential coefficient, by making h indefinitely small.

Form of the difference between two states of a function.

14. Let us resume the discussion of Equation (3).

$$y' - y = f(x + h) - f(x).$$

If h be made equal to 0, the first and second members will each reduce to 0. Therefore, if the second member be developed, and the like terms having contrary signs cancelled, each of the remaining terms will contain h; else, all the terms would not reduce to 0, when $h = 0$. Hence, the second member of Equation (3) is divisible by h. Dividing by h, we have,

$$\frac{y' - y}{h} = \frac{f(x + h) - fx}{h} \quad . \quad . \quad . \quad (4.)$$

If, now, we pass to the limiting ratio, by making h indefinitely small, the second member will become the tan a, a quantity independent of h (see Equation 5); hence, the first term in the development of the second member of Equation (3) contains h only in the first power, and the coefficient of this term is tan α, or the differential coefficient. Since all the other terms become 0, when $h = 0$, each of them must contain h to a higher power than the first.

If we designate by P, the differential coefficient of y, and by P' such a value that $P'h^2$ shall be equal to all the terms of the development of the second member of Equation (3), after the first, that equation may be written under the form

$$y' - y = Ph + P'h^2 \quad \ldots \quad \text{(6.)}$$

The differential coefficient, P, is independent of h, but will, in general, contain x; and when it does, it is a function of that variable: P', when not equal to 0, is a function of x and h.

Applications of the Formula.

1. If we have an expression of the form,

$$y = f(x) = ax,$$

we have the form or development of the second member.

If we give to x an increment h, we have,

$$y' = f(x + h) = \quad a(x \ + \ h) \ = ax \ + ah; \text{ hence,}$$

$$y' - y = f(x + h) - f(x) = ah; \text{ and}$$

$$\frac{y' - y}{h} = a; \quad \text{passing to consecutive values,}$$

$$\frac{dy}{dx} = a; \quad \text{and} \quad \frac{dy}{dx}dx = adx.$$

2. If we have a function of the form,

$$y = f(x) = ax^3,$$

we again have the form or development of the second member.

$$y' = f(x + h) = a(x + h)^3 = ax^3 + 3ax^2h + 3axh^2 + ah^3$$

$$y' - y = f(x + h) - f(x) = 3ax^2h + 3axh^2 + ah^3$$

$$\frac{y' - y}{h} = 3ax^2 + 3axh + ah^2 ;$$

passing to consecutive values, we have,

$$\frac{dy}{dx} = 3ax^2 ; \quad \text{and} \quad \frac{dy}{dx}dx = 3ax^2 dx.$$

In the first example $P = a$, and $P' = 0$. In the second, $P = 3ax^2$, and $P' = 3axh + ah^2$.

15. Equation (6) affords the means of determining the differential coefficient, and the differential of any function, whose form is developed in terms of the independent variable.

If we divide both members of Equation (6) by the increment h, and then pass to the limiting ratio, we have the differential coefficient. If we then multiply the differential coefficient by the differential of the independent variable, we have the differential of the function.

Equal functions have equal differentials

16. If two functions, u and v, dependent on the same variable x, are *equal* to each other, for all possible values of x, their differentials will also be equal.

For, x being the independent variable, we have (Art. **14**),

$$u' - u = Ph + P'h^2,$$
$$v' - v = Qh + Q'h^2,$$

in which P is the differential coefficient of u, regarded as a function of x, and Q the differential coefficient of v, regarded as a function of x.

But, since u' and v' are, by hypothesis, equal to each other, as well as u and v, we have,

$$Ph + P'h^2 = Qh + Q'h^2,$$

or, by dividing by h and passing to consecutive values,

$$P = Q,$$

hence,
$$\frac{du}{dx} = \frac{dv}{dx},$$

and,
$$\frac{du}{dx} dx = \frac{dv}{dx} dx,$$

that is, the differential of u is equal to the differential of v.

Converse not true.

17. The converse of this proposition is not generally true; that is,

If two differentials are equal to each other, we are not at liberty to conclude that the functions from which they were derived, are also equal.

For, let $\qquad u = v \pm A$ . . . . . . (1.)

in which A is a constant, and u and v both functions of x. Giving to x an increment h, we shall have,

$$u' = v' \pm A,$$

from which subtract Equation (1), and we obtain,

$$u' - u = v' - v,$$

and, by substituting for the difference between the two states of the function, we have,

$$Ph + P'h^2 = Qh + Q'h^2.$$

Dividing by h, and passing to consecutive values, we obtain,

$$P = Q; \quad \text{that is,} \quad \frac{du}{dx} = \frac{dv}{dx};$$

hence, $\qquad \dfrac{du}{dx} dx = \dfrac{dv}{dx} dx; \quad$ or, $\quad du = dv.$

Hence, the differentials of u and v are equal to each other, although v may be greater or less than u, by any constant quantity A; therefore,

Every constant quantity connected with a variable by the sign plus or minus, will disappear in the differentiation.

The reason of this is apparent; for, a constant does not increase or decrease with the variable; hence, there is no ultimate or last difference between two of its values; and this *ultimate or last difference* is the differential of a variable function. Hence, the differential of a constant quantity is equal to 0.

18. If we have a function of the form,

$$u = Av,$$

in which u and v are both functions of x, and give to x an increment h, we shall have,

$$u' - u = A(v' - v),$$

or, $\qquad Ph + P'h^2 = A(Qh + Q'h^2).$

Dividing by h, and passing to the consecutive values,

$$P = AQ,$$

or, $$P dx = A Q dx.$$

But, $du = P dx$,   and $dv = \dot{Q} dx$;

hence, $du = A dv$;   that is,

The differential of the product of a constant by a variable quantity, is equal to the constant multiplied by the differential of the variable.

Signs of the differential coefficient.

19. If u is any function of x, and we give to x an increment h, we have,

$$\frac{u' - u}{h} = P + P'h ;$$

and since h is positive, the sign of the first member will be positive when $u < u'$; that is, when u is an *increasing* function of x (Art. **7**). It will be negative when $u > u'$; that is, when u is a *decreasing* function of x. Passing to consecutive values, we have, under the first supposition,

$$\frac{du}{dx} = + P ; \text{ and}$$

$$\frac{du}{dx} = - P, \text{ under the second; hence,}$$

The differential coefficients have the same sign as the functions, when the functions are INCREASING, *and contrary signs, when they are* DECREASING.

If we multiply by dx, we obtain the differentials, which have the *same signs* as the differential coefficients.

Nature of a differential coefficient, and of a differential.

20. The method of treating the Differential Calculus, adopted in this treatise, is based on three hypotheses:

1st. That the independent variable changes uniformly:

2d. That in changing from one state of value to another, it passes through all the intermediate values; and,

3d. That any function dependent upon it, undergoes changes determined by the equation expressing the relations between them; and that such equation preserves the same general form.

If the independent variable changes uniformly, and assumes all possible values between the limits $x = a$, and $x = a'$, we have seen that the change cannot be denoted by a number. If, then, we denote this change by dx, we mean that dx is *smaller than any number;* hence,

$$\frac{1}{dx} = \infty \cdot$$

But, $$\frac{1}{dx} = \frac{dx}{dx^2} = \frac{dx^2}{dx^3} = \frac{dx^3}{dx^4}, \ \&c.$$

that is, any power of dx divided by a power of dx greater by 1, is *infinite;* hence, *any* power of dx is infinitely small, compared with the power next less. Hence, it follows:

1st. That the addition of dx to any number, can make no alteration in its value; and therefore, when connected with a numeral quantity by the sign $\pm$, may be omitted without error; thus,

$$3ax + dx = 3ax.$$

2d. Since dx^2 is infinitely small, compared with dx; that is, *infinitely less than dx*, we have,

$$5ax^2dx + dx^2 = 5ax^2dx;$$

and similarly for the higher powers of dx:

The quantities, dx, dx^2, dx^3, &c., are called *infinitely small quantities*, or INFINITESIMALS *of the first, second, and third orders:* from their law of formation, it follows that,

Every infinitely small quantity may be omitted without error when connected by the sign $\pm$ with any of a lower order.

Rate of change.

21. The measure of a quantity, great or small, is the number of times which it contains some other quantity of the same kind, regarded as a unit of measure.

In the Differential Calculus, dx, the differential of the independent variable, is the unit of measure. The *rate of change*, in the function y, is therefore expressed by $\dfrac{dy}{dx}$, and the actual change corresponding to dx, by

$$\frac{dy}{dx}\,dx = dy.$$

22. The equation of a straight line is,

$$y = ax + b.$$

If we take any point, as M, whose co-ordinates are y and x, and a second point N, whose co-ordinates are y', $x + h$, and we have,

$$y' - y = ah; \quad \text{or,} \quad \frac{y' - y}{h} = a \quad . \quad . \quad (1.)$$

that is,

$$\frac{NR}{MR} = \text{tangent } NMR = a\,;$$

and, passing to the consecutive
values,

$$\frac{dy}{dx} = \text{tangent } \alpha = a \quad \ldots \ldots \quad (2.)$$

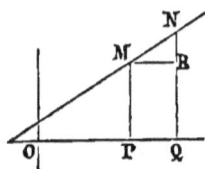

The differential coefficient $\frac{dy}{dx}$, measures the *rate of increase* of the ordinate y, when x receives the increment dx; and since this value is independent of x, the rate will be the same for every point of the line; that is, the *rate of ascension* of the line from the axis of abscissas, is the same at every point. And since,

$$\frac{dy}{dx}\,dx = dy = adx,$$

the *change* in the value of the ordinate will be *uniform*, for uniform changes in the abscissa.

23. Let us examine an equation,

$$y = f(x) \quad . \quad . \quad (1.)$$

not of the first degree.

Let us suppose the curve AMN to be such that the abscissas and ordinates of its different points shall correspond to all possible relations between y and x, in Equation (1).

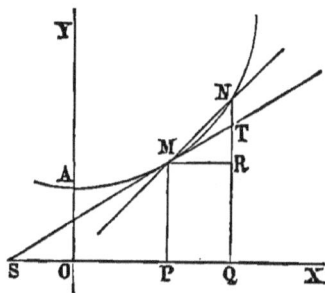

We have seen (Art. **13**) that,

$$\frac{dy}{dx} = \tan TSP = \tan \alpha\,; \text{ hence,}$$

the *rate of increase* of the function, or the *ascension* of the curve at any point, is equal to the tangent of the angle which the tangent line makes with the axis of abscissas. We also see, that this value of the tangent of α, will vary with the position of the point M; hence it s a function of x; therefore,

In every equation, not of the first degree, the differential coefficient is a function of the independent variable.

1. We have seen, that when the points M and N are consecutive, the secant line, MN, becomes the tangent line, TMS (Art. **13**). The line MR is then denoted by dx, and RN or RT, (for the points N and T then coincide), by dy. If we give to the new abscissa, $x + dx$, an additional increment dx, and suppose the corresponding ordinate, $y + dy$, to receive the *same increment as before*, viz.: dy, the extremity of the last ordinate will not fall on the curve, but on the tangent line, since the triangles thus formed are similar; hence,

If a function be supposed to increase uniformly from any assumed value, the differential coefficient will be constant, and equal to any increment of the function divided by the corresponding increment of the variable.

Nature of the Differential Calculus.

24. In every operation of the Differential Calculus, one of two things is always proposed, and sometimes both:

1st. To find the *rate of change* in any variable function when it *begins* to change from any assigned value.

2d. To find the difference between any two consecutive values of the function. This difference is the *actual change* in the function, produced by the smallest change which takes place in the independent variable.

The use of the independent variable is to furnish a unit of measure for the increment of the function, and thus to determine its *rate of change*, as it passes through all its states of value. This ratio can generally be expressed in numbers, either exactly or approximatively.

25. The increment of the function, corresponding to the smallest increment of the variable, being the difference between any two of its consecutive values, is a quantity of the *same kind* as the function, and differs from it only in this: that it is *too small to be expressed by numbers*. The differential of a quantity, therefore, is merely an *element* of that quantity; that is, it is the change which takes place when the quantity *begins* to increase or decrease, from any assumed value. When we find this element, we have the differential of the function ; and by dividing by dx, we have the differential coefficient. Hence,

All the operations of the Differential Calculus comprise but two objects :

1. *To find the rate of change in a function, when it passes from one state of value to another, consecutive with it.*

2. *To find the actual change in the function.*

The rate of change is the *differential coefficient*, and the actual change, the *differential*.

SECTION II.

Differential of sum or difference of Functions.

26. Let u be a function of the algebraic sum of several variable quantities, of the form,

$$u = y + z - w = f(x),$$

in which y, z, and w, are functions of the independent variable x.

If we give to x an increment h, we shall have,

$$u' - u = (y' - y) + (z' - z) - (w' - w);$$

hence (Art. **14**),

$$u' - u = (Ph + P'h^2) + (Qh + Q'h^2) - (Lh + L'h^2),$$

or, $\quad \dfrac{u' - u}{h} = (P + P'h) + (Q + Q'h) - (L + L'h),$

and by passing to consecutive values,

$$\frac{du}{dx} = P + Q - L;$$

multiplying both members by dx, we have,

$$\frac{du}{dx} dx = Pdx + Qdx - Ldx.$$

But as P, Q, and L, are the differential coefficients

of y, z, and w, each regarded as a function of x; hence,

$$\frac{du}{dx}dx = \frac{dy}{dx}dx + \frac{dy}{dx}dx - \frac{dw}{dx}dx; \text{ that is,}$$

The differential of the sum or difference of any number of functions, dependent on the same variable, is equal to the sum or difference of their differentials taken separately

Differential of a product.

27. Let u and v denote any two functions, x the independent variable, and h its increment; we shall then have,

$$u' = u + Ph + P'h^2, \text{ and}$$
$$v' = v + Qh + Q'h^2,$$

and, multiplying,

$$u'v' = (u + Ph + P'h^2)(v + Qh + Q'h^2)$$
$$= uv + vPh + uQh + PQh^2 + \&c.;$$

hence,

$$\frac{u'v' - uv}{h} = vP + uQ + \text{terms containing } h, h^2, \text{ and } h^3.$$

If now we pass to consecutive values, we have,

$$\frac{d(uv)}{dx} = vP + uQ;$$

therefore, $d(uv) = vPdx + uQdx = vdu + udv$; hence,

The differential of the product of two functions dependent on the same variable, is equal to the sum of the products obtained by multiplying each by the differential of the other.

1. If we divide by uv, we have,

$$\frac{d(uv)}{uv} = \frac{du}{u} + \frac{dv}{v} \cdot \cdot \cdot \cdot \cdot (1.)$$

that is,

The differential of the product of two functions, divided by the product, is equal to the sum of the quotients obtained by dividing the differential of each by its function.

28. We can easily determine, from the last formula, the differential of the product of any number of functions. For, put $v = ts$, then,

$$\frac{dv}{v} = \frac{d(ts)}{ts} = \frac{dt}{t} + \frac{ds}{s} \cdot \cdot \cdot \cdot \cdot (2.)$$

and by substituting ts for v, in Equation (1), we have,

$$\frac{d(uts)}{uts} = \frac{du}{u} + \frac{dt}{t} + \frac{ds}{s};$$

and in a similar manner we should find,

$$\frac{d(utsr\ldots)}{utsr\ldots} = \frac{du}{u} + \frac{dt}{t} + \frac{ds}{s} + \frac{dr}{r} \ldots \text{&c.}$$

If, in the equation,

$$\frac{d(uts)}{uts} = \frac{du}{u} + \frac{dt}{t} + \frac{ds}{s},$$

we multiply by the denominator of the first member, we shall have,

$$d(uts) = tsdu + usdt + utds; \quad \text{hence,}$$

The differential of the product of any number of func tions, is equal to the sum of the products which arise by multiplying the differential of each function by the product of all the others.

Differentials of Fractions.

29. To obtain the differential of any fraction of the form, $\dfrac{u}{v}$.

Put, $\dfrac{u}{v} = t,$ then, $u = tv.$

Differentiating both members, we have,

$$du = vdt + tdv;$$

finding the value of dt, and substituting for t its value $\dfrac{u}{v}$, we obtain,

$$dt = \frac{du}{v} - \frac{udv}{v^2},$$

or, by reducing to a common denominator,

$$dt = \frac{vdu - udv}{v^2}; \text{ hence,}$$

The differential of a fraction is equal to the denominator into the differential of the numerator, minus the numerator into the differential of the denominator, divided by the square of the denominator.

1. If the denominator is constant, $dv = 0$, and we have,

$$dt = \frac{vdu}{v^2} = \frac{du}{v}.$$

2. If the numerator is constant, $du = 0$, and we have,

$$dt = -\frac{u\,dv}{v^2};$$

and under this supposition, t is a decreasing function of v (Art. **7**); hence, its differential coefficient should be negative (Art. **19**).

Differentials of Powers.

29. * To find the differential of any power of a function. First, take any function u^n, in which n is a positive whole number. This function may be considered as composed of n factors, each equal to u. Hence (Art. **27**),

$$\frac{d(u^n)}{u^n} = \frac{d(uuuu\ldots.)}{(uuuu\ldots.)} = \frac{du}{u} + \frac{du}{u} + \frac{du}{u} + \frac{du}{u} + \ldots..$$

But as there are n equal factors in the numerator of the first member, there will be n equal terms in the second;

hence, $$\frac{d(u^n)}{u^n} = \frac{n\,du}{u};$$

therefore, $$d(u^n) = nu^{n-1}\,du.$$

1. If n is fractional, denote it by $\frac{r}{s}$, and make,

$$v = u^{\frac{r}{s}}, \qquad \text{whence,} \qquad v^s = u^r;$$

and since r and s are entire numbers, we shall have,

$$sv^{s-1}\,dv = ru^{r-1}\,du;$$

from which we find,

$$dv = \frac{ru^{r-1}}{sv^{s-1}}\,du = \frac{ru^{r-1}}{su^{\frac{r}{s}(s-1)}}\,du;$$

or, by reducing,

$$dv = \frac{r}{s} u^{\frac{r}{s}-1} du;$$

which is obtained directly from the function,

$$d(u^n) = nu^{n-1} du,$$

by changing the exponent n to $\frac{r}{s}$.

2. If the fractional exponent is one-half, the function becomes a radical of the second degree. We will give a specific rule for this class of functions.

Let $\qquad v = u^{\frac{1}{2}}, \qquad$ or, $\qquad v = \sqrt{u};$

then, $\qquad dv = \frac{1}{2}u^{\frac{1}{2}-1}du = \frac{1}{2}u^{-\frac{1}{2}}du = \frac{du}{2\sqrt{u}};$

that is,

The differential of a radical of the second degree, is equal to the differential of the quantity under the sign divided by twice the radical.

3. Finally, if n is negative, we shall have,

$$u^{-n} = \frac{1}{u^n},$$

from which we have (Art. **29**),

$$d(u^{-n}) = d\left(\frac{1}{u^n}\right) = \frac{-d(u^n)}{u^{2n}} = \frac{-nu^{n-1}du}{u^{2n}};$$

and, by reducing,

$$d(u^{-n}) = -nu^{-n-1}du; \quad \text{hence,}$$

The differential of any power of a function, is equal to the exponent multiplied by the function raised to a power less one, multiplied by the differential of the function.

Formulas for differentiating Algebraic Functions.

1. $d(a)$ $= 0$ . . . . (Art. **17.**)

2. $d(ax)$ $= adx$ . . . . (Art. **18.**)

3. $d(x + y) = dx + dy$ . . . . . . (Art. **26.**)

4. $d(x - y) = dx - dy$ . . . . . . (Art. **26.**)

5. $d(xy)$ $= xdy + ydx$ . . . . . (Art. **27.**)

6. $d\left(\dfrac{x}{y}\right)$ $= \dfrac{ydx - xdy}{y^2}$ . . . . . (Art. **29.**)

7. $d(x^m)$ $= mx^{m-1}dx$ . . . . . . (Art. **30.**)

8. $d(\sqrt{x})$ $= \dfrac{dx}{2\sqrt{x}}$ . . . . (Art. **30**—2.)

9. $d(x^{-\frac{r}{s}})$ $= -\dfrac{r}{s}x^{-\frac{r}{s}-1}dx$ . . . . (Art. **30**—3.)

EXAMPLES.

Find the differentials of the following functions:

1. $u = ax - y.$ $du = adx - dy.$

2. $u = a^2x^2 + z.$ $du = 2a^2xdx + dz.$

3. $u = bx^2 - y^3 + a.$ $du = 2bxdx - 3y^2dy.$

4. $u = ax^2 - bx^3 + x.$ $du = (2ax - 3bx^2 + 1)dx.$

5. $u = cy^2 - x^2 + ay^2.$ $du = 2[(c + a)ydy - xdx.]$

6. $u = xyz.$ $du = yzdx + xzdy + xydz.$

7. $u = y^2 - a^2 - 8az^5.$ $du = 2(ydy - 20az^4dz.)$

8. $u = 3a^5x^n.$ $du = 3na^5x^{n-1}dx.$

9. $u = -2ax^{-5} - 5 + 4b^2x^3.$ $du = 2\left(\dfrac{5a}{x^6} + 6b^2x^2\right)dx.$

10. $u = 5x^5 - 2ay - b^2.$ $du = 25x^4dx - 2ady.$

11. $u = x^n - x^3 + 46.$ $du = (nx^{n-1} - 3x^2)dx.$

12. $u = ax(x^2 + 3b).$ $du = 3a(x^2 + b)dx.$

13. $u = (x^2 + a)(x - a).$ $du = (3x^2 - 2ax + a)dx.$

14. $u = x^2y^2z^3.$ $du = 2xy^2z^3dx + 2x^2z^3ydy + 3x^2y^2z^2dz.$

15. $u = ax^2(x^3 + a).$ $du = ax(5x^3 + 2a)dx.$

16. $u = \dfrac{x}{y}.$ $du = \dfrac{ydx - xdy}{y^2}.$

17. $u = \dfrac{a}{b - 2y^2}.$ $du = \dfrac{4aydy}{(b - 2y^2)^2}.$

18. $u = \dfrac{1}{x}.$ $du = \dfrac{-dx}{x^2}.$

19. $u = x^{-n} = \dfrac{1}{x^n}.$ $du = \dfrac{-ndx}{x^{n+1}}.$

20. Find the differential of u in the equation,

$$u = \sqrt{a^2 - x^2}.$$

Put, $a^2 - x^2 = y$; then, $u = \sqrt{y}$; and (Art. 30—2),

$$du = \dfrac{dy}{2\sqrt{y}}.$$

But, $dy = -2xdx$; then, substituting for y and dy, their values, we have,

$$du = \frac{-2xdx}{2\sqrt{a^2-x^2}} = \frac{-xdx}{\sqrt{a^2-x^2}}.$$

21. $u = \sqrt{2ax + x^2}$ $du = \dfrac{(a+x)dx}{\sqrt{2ax+x^2}}.$

22. $u = \dfrac{1}{\sqrt{1-x^2}}.$ $du = \dfrac{xdx}{(1-x^2)^{\frac{3}{2}}}.$

23. $u = \dfrac{x}{x+\sqrt{1-x^2}}.$ $du = \dfrac{dx}{\sqrt{1-x^2}\left(x+\sqrt{1-x^2}\right)^2}.$

24. $u = \left(a + \sqrt{x}\right)^3.$ $du = \dfrac{3\left(a+\sqrt{x}\right)^2 dx}{2\sqrt{x}}.$

25. $u = \dfrac{a^2 - x^2}{a^4 + a^2x^2 + x^4}.$

$$du = \frac{(a^4+a^2x^2+x^4)d(a^2-x^2) - (a^2-x^2)d(a^4+a^2x^2+x^4)}{(a^4+a^2x^2+x^4)^2},$$

or, $du = \dfrac{-2x(2a^4 + 2a^2x^2 - x^4)dx}{(a^4+a^2x^2+x^4)^2}.$

26. $u = \sqrt{a^2 + x^2} \times \sqrt{b^2 + y^2}.$

$$du = \frac{(b^2+y^2)xdx + (a^2+x^2)ydy}{\sqrt{a^2+x^2}\sqrt{b^2+y^2}}.$$

27. $u = \dfrac{x^n}{(1+x)^n}.$ $du = \dfrac{nx^{n-1}dx}{(1+x)^{n+1}}.$

28. $u = \dfrac{1+x^2}{1-x^2}.$ $du = \dfrac{4xdx}{(1-x^2)^2}.$

29. $u = \dfrac{x + y}{z^3}$. $du = \dfrac{z(dx + dy) - (x + y)3dz}{z^4}$.

30. $u = \dfrac{\sqrt{1 + x} + \sqrt{1 - x}}{\sqrt{1 + x} - \sqrt{1 - x}}$.

$$du = -\frac{(1 + \sqrt{1 - x^2})dx}{x^2\sqrt{1 - x^2}}.$$

Differential of a particular binomial.

30.—1. Let $u = (a + bx^n)^m$.

Put $a + bx^n = y$; then, $u = y^m$; and (Art. **30**),

$$du = my^{m-1}dy.$$

But, from the first equation,

$$dy = nbx^{n-1}dx;$$

substituting for y and dy their values, we have,

$$du = mnb(a + bx^n)^{m-1}x^{n-1}dx;$$

that is, to find the differential of a binominal function of this form,

Multiply the exponent of the parenthesis, into the exponent of the variable within the parenthesis, into the coefficient of the variable, into the binomial raised to a power less 1, into the variable within the parenthesis raised to a power less 1, into the differential of the variable.

Rate of change of the Function.

31. What is the rate of change in the area of a square, when the side is denoted by the independent variable?

We have seen (Art. **21**) that the differential coefficient, $\dfrac{du}{dx}$, denotes the *rate* of change in the function u, corresponding to the change dx, in the value of x; and that in all equations, except those of the first degree, this rate will be *variable*, and a function of x (Art. **23**).

Let x denote the side of a square, and u its area; then,

$$u = x^2, \quad \text{and} \quad \frac{du}{dx} = 2x;$$

hence, *the rate of change in the area of a square is equal to twice its side;* that is, if the side of a square is denoted by 1, the rate of change in the area will be denoted by 2; if the edge is denoted by 5, the rate of change will be 10; and similarly for other numbers.

2. What is the rate of change in the volume of a cube, when its edge is the independent variable?

Let x denote the edge of a cube, and u its volume; then,

$$u = x^3, \quad \text{and} \quad \frac{du}{dx} = 3x^2;$$

hence, the *rate* of change in the volume, *is three times the square of its edge.* If the edge is denoted by 1, the rate of change in the volume will be denoted by 3; if the edge is denoted by 2, the rate of change will be 12; if 3, the rate will be 27; and similarly, when the edge is denoted by other numbers.

Find the rates of change in the following functions:

3. $u = 8x^4 - 3x^2 - 5x + a.$ $A.\ 32x^3 - 6x - 5.$

What will express the rate for

$$x = 1, \quad x = 2, \quad x = 3?$$

4. $u = (x^3 + a)(3x^2 + b).$ $A.\ 15x^4 + 3x^2 b + 6ax.$

Find the rate for,

$$x = 1, \quad x = 2.$$

5. $u = \dfrac{1}{1-x}$.   *A.* $+\dfrac{1}{(1-x)^2}$.

What is the rate for, $x = 0$, $x = 4$, $x = -1$?

6. $u = (ax + x^2)^2$.   *A.* $2(ax + x^2)(a + 2x)$.

What is the rate for, $x = 0$, $x = 1$, $x = 3$?

7. $u = \dfrac{x}{x + \sqrt{1 - x^2}}$.   *A.* $\dfrac{1}{\sqrt{1 - x^2}(x + \sqrt{1 - x^2})^2}$.

What is the rate for, $x = 0$, $x = 1$?

Hence, to find the rate of change for a given value of the variable : *Find the differential coefficient, and substitute the value of the variable in the second member of the equation.*

Partial Differentials.

32. If we have a function of the form,

$$u = f(x, y) \quad \ldots \ldots \quad (1.)$$

the equation denotes that u is a function of the two variables, x and y. If we suppose either of these, as y, to remain constant, and x to vary, we shall have,

$$\frac{du}{dx} = f'(x, y) . \quad \ldots \ldots \quad (2.)$$

if we suppose x to remain constant, and y to vary, we shall have,

$$\frac{du}{dy} = f''(x, y) \quad \ldots \ldots \quad (3.)$$

The differential coefficients which are obtained under these suppositions, are called *partial differential coefficients*.

The first is the partial differential coefficient with respect to x, and the second with respect to y.

33. If we multiply both members of Equation (2) by dx, and both members of Equation (3) by dy, we obtain,

$$\frac{du}{dx}\,dx = f'(x, y)dx, \quad \text{and} \quad \frac{du}{dy}\,dy = f''(x, y)dy.$$

The expressions,

$$\frac{du}{dx}\,dx, \qquad \frac{du}{dy}\,dy,$$

are called, *partial differentials;* the first a partial differential with respect to x, and the second a partial differential with respect to y; hence,

A PARTIAL DIFFERENTIAL COEFFICIENT *is the differential coefficient of a function of two or more variables, under the supposition that only one of them has changed its value;* and,

A PARTIAL DIFFERENTIAL *is the differential of a function of two or more variables, under the supposition that only one of them has changed its value.*

If we suppose both the variables to undergo a change at the same time, the corresponding change which takes place in u, is called, the *total differential.* If we extend this definition to any number of variables, and assume what may be rigorously proved, viz. :

That the total differential of a function of any number of variables is equal to the sum of the partial differentials,

we have a general formula applicable to every funo-
tion of two or more variables.

EXAMPLES.

1. Let $u = x^2 + y^3 - z$; then,

$$\frac{du}{dx}\, dx = 2xdx, \qquad \text{1st partial differential},$$

$$\frac{du}{dy}\, dy = 3y^2dy, \qquad \text{2d} \quad \text{``} \qquad \text{``}$$

$$\frac{du}{dz}\, dz = -\, dz, \qquad \text{3d} \quad \text{``} \qquad \text{``}$$

hence, $du = 2xdx + 3y^2dy - dz.$

2. Let $u = xy$; then,

$$\frac{du}{dx}\, dx = ydx,$$

$$\frac{du}{dy}\, dy = xdy\,;$$

hence, $du = ydx + xdy.$

3. Let $u = x^m y^n$; then,

$$\frac{du}{dx}\, dx = mx^{m-1}y^n dx,$$

$$\frac{du}{dy}\, dy = ny^{n-1}x^m dy\,; \quad \text{hence,}$$

$$du = mx^{m-1}y^n dx + ny^{n-1}x^m dy = x^{m-1}y^{n-1}(mydx + nxdy).$$

4. Let $u = \dfrac{x}{y}$; then,

$$\frac{du}{dx}\,dx = \frac{dx}{y},$$

$$\frac{du}{dy}\,dy = -\frac{x\,dy}{y^2};$$

hence, $\qquad du = \dfrac{y\,dx - x\,dy}{y^2}.$

5. Let $u = \dfrac{ay}{\sqrt{x^2 + y^2}} = ay(x^2 + y^2)^{-\frac{1}{2}}$; then,

$$\frac{du}{dx}\,dx = -\frac{ayx\,dx}{(x^2 + y^2)^{\frac{3}{2}}},$$

$$\frac{du}{dy}\,dy = \frac{a\,dy}{(x^2 + y^2)^{\frac{1}{2}}} - \frac{ay^2\,dy}{(x^2 + y^2)^{\frac{3}{2}}};$$

hence, $\qquad du = -\dfrac{ayx\,dx - ax^2\,dy}{(x^2 + y^2)^{\frac{3}{2}}}.$

6. Let $u = xyzt$; then,

$$du = yzt\,dx + xzt\,dy + xyt\,dz + xyz\,dt.$$

SECTION III.

34. AN INTEGRAL is a functional expression, either algebraic or transcendental, derived from a differential.

DIFFERENTIATION and INTEGRATION are terms denoting operations the exact converse of each other.

DIFFERENTIATION is the operation of finding the differential function from the primitive function.

INTEGRATION is the operation of finding the primitive function from the differential function.

Rules have been found for the differentiation of every form which a function can assume. Hence, in the Differential Calculus, no case can occur to which a known rule is not applicable. In the Integral Calculus it is quite otherwise.

In returning from a known differential to the integral from which it may have been derived, we *compare the differential expression with other expressions which are known to be differentials of given functions*, and thus arrive at the form of the integral, or primitive function. The main operations, therefore, of the Integral Calculus, consist in *transforming given differential expressions into others which are equivalent to them*, and which are differentials of known functions ; and thus deducing formulas applicable to all similar forms.

The integration is indicated by placing the sign $\int$

before the expression to be integrated. It is equivalent
to "integral of"; thus,

$$\int 2x\,dx = x^2,$$

is read: "Integral of $2x\,dx$, is equal to x^2."

Integration of Monomials.

35. The differential of every expression of the form,

$$u = x^m, \quad \text{is} \quad du = mx^{m-1}dx \quad \text{(Art. 30)},$$

which has been found by *multiplying the exponent into
the variable raised to a power less one, into the differential of the variable.*

If, then, we have a differential expression, of the form,

$$mx^{m-1}dx, \quad \text{or,} \quad x^m dx,$$

we can find its integral by reversing the above rule; that
is, to find the integral of such an expression,

*Add 1 to the exponent of the variable, and then divide
by the new exponent into the differential of the variable.**

Find the integrals of the following differential expressions ·

1. If $du = 2x\,dx$, $\qquad \displaystyle\int du = \frac{2x^2 dx}{2 \times dx} = x^2.$

2. If $du = 3x^2 dx$, $\qquad \displaystyle\int du = \frac{3x^3 dx}{3 \times dx} = x^3.$

---

* This rule applies to every case of a differential monomial of the
form, $Ax^m dx$, except that in which m is -1 (Art. **90**).

3. If $du = x^m dx,$ $\int du = \dfrac{x^{m+1} dx}{(m+1) dx} = \dfrac{x^{m+1}}{m+1}.$

4. If $du = x^{-3} dx,$ $\int du = \dfrac{x^{-3+1} dx}{-2 dx} = -\dfrac{1}{2x^2}.$

5. If $du = x^3 \sqrt{x}\, dx,$ $\int du = \int x^{\frac{7}{2}} dx = \dfrac{2}{9} x^4 \sqrt{x}.$

36. We have seen, that the differential of the product of a constant by a variable, is equal to the constant multiplied by the differential of the variable (Art. **18**). Hence, *the integral of the product of a constant by a differential, is equal to the constant multiplied by the integral of the differential;* that is,

$$\int a x^m dx = a \int x^m dx = a \frac{1}{m+1} x^{m+1}.$$

Hence, *if the expression to be integrated has one or more constant factors, they should, at once, be placed as factors, without the sign of the integral.*

37. It has been shown that the differential of the sum or difference of any number of variables is equal to the sum or difference of their differentials (Art. **26**). Hence, if we have a differential expression of the form,

$$du = 2ax^2 dx - by dy - z^2 dz; \text{ we may write,}$$

$$\int du = 2a \int x^2 dx - b \int y dy - \int z^2 dz; \text{ or,}$$

$$\int du = \frac{2}{3} a x^3 - \frac{b}{2} y^2 - \frac{z^3}{3}; \text{ that is,}$$

The integral of the algebraic sum of any number of differentials is equal to the algebraic sum of their integrals.

Correction—Indefinite—Particular—and Definite Integrals.

38. It has been shown that every constant quantity connected with a variable by the sign plus or minus, disappears in the differentiation (Art. **17**); that is,

$$d(a + x^m) = dx^m = mx^{m-1} dx.$$

Hence, the same differential may have several integral functions differing from each other by a constant term. Therefore, in passing from a differential to an integral. expression, we must annex to the first integral obtained, a constant term, to compensate for the constant term which may have been lost in the differentiation.

For example, it has been shown in Art. (**22**), that,

$$\frac{dy}{dx} = a, \quad \text{or,} \quad dy = adx,$$

is the differential equation of every straight line which makes with the axis of abscissas an angle whose tangent is a. Integrating this expression, we have,

$$\int dy = a\int dx . \quad . \quad . \quad . \quad . \quad (1.)$$

or, $y = ax$;

or, finally, $y = ax + C$ . . . . . (2.)

If, now, the required line is to pass through the origin of co-ordinates, we shall have, for

$$x = 0, \quad y = 0, \quad \text{and consequently,} \quad C = 0.$$

But if it be required that the line shall intersect the

axis of Y at a distance from the origin equal to $+ b$, we shall have, for

$$x = 0, \quad y = + b, \quad \text{and} \quad \text{consequently,} \quad C = + b;$$

and the true integral will be,

$$y = ax + b \ . \ . \ . \ . \ . \ . \ (3.)$$

If, on the contrary, it were required that the right line should intersect the axis of ordinates below the origin, we should have, for

$$x = 0, \quad y = - b, \quad \text{and consequently,} \quad C = - b;$$

and the true integral would be,

$$y = ax - b \ . \ . \ . \ . \ . \ . \ (4.)$$

The constant C, which is added to the first integral, *must have such a value as to render the functional equation true for every possible value that may be attributed to the variable.* Hence, after having found the first integral equation, and added the constant C, *if we then make the variable equal to zero*, the value which the function assumes will be the true value of C.

1. An *indefinite* integral is the first integral obtained, before the value of the constant C is determined.

2. A *particular* integral is the integral after the value of C has been found.

3. A *definite* integral is the integral corresponding to a given value of the variable.

Thus, Equation (2) is an indefinite integral, because, so long as C is undetermined, it will be the equation of t system of parallel straight lines. Equations (3) and (4) are particular integrals, because each belongs to a particular line.

Origin of the Integral.

39. THE ORIGIN of an integral function is its zero value. The value of the variable corresponding to the origin of the integral, is found by placing the second member of the equation expressing the particular integral, equal to zero, and finding therefrom the value of the variable. Thus, if in Equation (3), we make $y = 0$, we have,

$$ax + b = 0, \quad \text{and} \quad x = -\frac{b}{a},$$

which shows that the origin of the function y (that is $y = 0$), is on the side of negative abscissas, and at a distance from the origin equal to $-\frac{b}{a}$. In Equation (4), it is at a point whose abscissa is $\frac{b}{a}$.

Integration between limits.

40. Having found the indefinite integral, and the particular integral, the next step is to find the definite integral; and then, the *definite* integral between given limits of the variable.

Let us take the particular integral found in Equation (3),

$$y = ax + b.$$

If it is required to find the value of the function y, for a given value of the variable x, as, $x = x'$, y will become a constant for this value, and we shall have,

$$y' = ax' + b \quad . \quad . \quad . \quad . \quad (5.)$$

which is a *definite* integral.

If we wish the value of the function corresponding to a second abscissa, $x = x''$, we shall have,

$$y'' = ax'' + b \quad \cdots \cdots \quad (6.)$$

If we subtract Equation (5) from Equation (6), we have,

$$y'' - y' = a(x'' - x') \quad \cdots \cdots \quad (7.)$$

which is the definite integral of y, taken between the limits, $x = x'$, and $x = x''$.

If, $x' = OP$, and $x'' = OQ$; then,

$y' = PM$, and $y'' = QN$; hence,

$y'' - y' = a(x'' - x') = NR$;

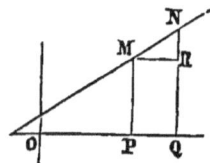

Therefore: *The integral of a function, taken between two limits, indicated by given values of x, is equal to the difference of the definite integrals corresponding to those limits.*

Let us now explain the *language* employed to express these relations. The modified form of Equation (1),

$$\int (dy)_{x=x'} = a\int dx,$$

is read: "Integral of y, when x is equal to x';" and

$$\int (dy)_{x=x''} = a\int dx,$$

is read: "Integral of y, when x is equal to x'';" and

$$\int_{x'}^{x''} (dy) = a\int dx,$$

is read: Integral of the differential of y, taken between the limits, x' and x''; the least limit, or the limit corresponding to the subtractive integral, being placed below.

EXAMPLE.

1. What is the integral of $du = 9x^2dx$, between the limits $x = 1$, and $x = 3$, if in the primitive function u reduces to 81, when $x = 0$.

$$\int du = \int 9x^2dx = 3x^3 + C \; ; \quad \text{hence,}$$

$$\int du = 3x^3 + C.$$

But from the primitive function, $u = 81$, when $x = 0$; hence, $C = 81$, and,

$$\int du = 3x^3 + 81 \quad . \quad . \quad . \quad . \quad (1.)$$

$$\int (du)_{x=1} = 3 + 81 = 84 \quad . \quad . \quad (2.)$$

$$\int (du)_{x=3} = 81 + 81 = 162 \quad . \quad . \quad (3.)$$

$$\int_1^3 (du) = 162 - 84 = 78 \quad . \quad . \quad (4.)$$

What is the value of the variable corresponding to the origin of the integral (Art. **39**)?

Making the second member of Equation (1) equal 0,

$$3x^3 + 81 = 0, \quad \text{or,} \quad x = -3.$$

Integration of particular binomials.

41. To integrate a differential of the form (Art. **30**),

$$du = (a + bx^n)^m x^{n-1}dx. \quad . \quad . \quad . \quad (1.)$$

The characteristic of this form is, that *the exponent of the variable without the parenthesis is less by* 1 *than the* exponent of the variable within.

Put, $(a + bx^n) = z$;   then, $(a + bx^n)^m = z^m$; and,

$$nbx^{n-1}dx = dz;   \text{whence,}   x^{n-1}dx = \frac{dz}{nb}; \text{hence,}$$

$$\int du = \int (a + bx^n)^m x^{n-1}dx = \int \frac{z^m dz}{nb} = \frac{z^{m+1}}{(m+1)nb};$$

and consequently,

$$u = \frac{(a + bx^n)^{m+1}}{(m+1)nb} + C.$$

Hence, to find the integral of the above form,

1. *If there is a constant factor, place it without the sign of the integral, and omit the power of the variable without the parenthesis and the differential:*

2. *Augment the exponent of the parenthesis by* 1, *and then divide this quantity, with its exponent so increased, by the exponent of the parenthesis, into the exponent of the variable within the parenthesis, into the coefficient of the variable.*

EXAMPLES.

1. $$\int (a + 3x^2)^3 x dx = \frac{(a + 3x^2)^4}{4.2.3} + C; \text{and}$$

2. $$\int m(a + bx^2)^{\frac{1}{2}} x dx = \frac{m}{3b}(a + bx^2)^{\frac{3}{2}} + C.$$

3. $$\int mn(a - 4cx^4)^{\frac{3}{2}} x^3 dx = -\frac{mn}{40c}(a - 4cx^4)^{\frac{5}{2}} + C.$$

Integration by Series.

42. The approximate integral of any function of the form,

$$du = Xdx,$$

may be found, when X is such a function of x, that it can be developed into a series. Having made the development of the function X, in the powers of x, by the Binomial Formula, we multiply each term by dx, and then integrate the terms separately. When the series is converging, we readily find the approximate value of the function for any assumed value of the variable.

EXAMPLE.

1. Find the approximate integral of,

$$\int du = \int \frac{dx}{\sqrt{1 - x^2}} = \int (1 - x^2)^{-\frac{1}{2}} dx,^*$$

in which, $\qquad X = (1 - x^2)^{-\frac{1}{2}}.$

Developing, $(1 - x^2)^{-\frac{1}{2}}$, by the binomial formula,†

$$(1 - x^2)^{-\frac{1}{2}} = 1 + \frac{1}{2}x^2 + \frac{1}{2}\cdot\frac{3}{4}x^4 + \frac{1}{2}\cdot\frac{3}{4}\cdot\frac{5}{6}x^6 + \&c.;$$

multiplying by dx, and integrating, we obtain,

$$\int du = x + \frac{1}{2}\frac{x^3}{3} + \frac{1}{2}\cdot\frac{3}{4}\frac{x^5}{5} + \frac{1}{2}\cdot\frac{3}{4}\cdot\frac{5}{6}\frac{x^7}{7} + \&c.$$

---

* Bourdon, Art. **166.** University, Art. **32.**
† Bourdon, Art. **135.** University, Art. **104.**

from which we obtain an approximate value of u, cor-responding to any value we may give to x.

APPLICATIONS TO GEOMETRICAL MAGNITUDES.

Equations of Tangents and Normals.

43. We have seen, that if x and y denote the co-ordinates of every point of a curve, $\dfrac{dy}{dx}$ will denote the tangent of the angle which the tangent line makes with the axis of abscissas (Art. **13**). This value of $\dfrac{dy}{dx}$ was found under the supposition that the second secant point became *consecutive with the first;* hence,

Any two consecutive points, must, at the same time, be in the chord, the curve, and the tangent.

Denote the co-ordinates of the point of tangency, in any curve, by x'' and y''. If through this point we draw any secant line, its equation will be of the form,

$$y - y'' = a(x - x'').^*$$

If the second point of secancy becomes consecutive with the first, we shall have (Art. **13**),

$$a = \frac{dy''}{dx''};$$

hence, the equation of the tangent line is,

$$y - y'' = \frac{dy''}{dx''}(x - v'') \ . \ . \ . \ . \ (1.)$$

* Bk. I. Art. **20.**

If, in the equation of any curve, we find the value of $\frac{dy''}{dx''}$, and substitute that value in Equation (1), the equation will then denote the tangent to that curve.

1. By differentiating the equation of the circle,

$$x^2 + y^2 = R^2, \qquad \text{or,} \qquad x''^2 + y''^2 = R^2,$$

we have,
$$\frac{dy''}{dx''} = - \frac{x''}{y''}; *$$

hence,
$$y - y'' = - \frac{x''}{y''}(x - x'');$$

or, by reducing, $yy'' + xx'' = R^2$.   .

2. By differentiating the equation of the ellipse, we have,

$$\frac{dy''}{dx''} = - \frac{B^2 x''}{A^2 y''} \cdot \dagger$$

3. By differentiating the equation of the parabola, we have,

$$\frac{dy''}{dx''} = \frac{p}{y''} \cdot \ddagger$$

4. By differentiating the equation of the hyperbola, we have,

$$\frac{dy''}{dx''} = \frac{B^2 x''}{A^2 y''}.$$

Substituting these values, in succession, in Equation (1), and reducing, we shall find the equation of the tangent line to each curve.

---

* Bk. II. Art. **8**.   † Bk. III. Art. **14**.   ‡ Bk. IV. Art. **8**.

44. The equation of the normal is of the form,

$$y - y'' = a'(x - x'') \quad \ldots \quad \ldots \quad (1.)$$

But since the normal is perpendicular to the tangent, at the point of contact,

$$1 + aa' = 0,^* \quad \text{or,} \quad a' = -\frac{1}{a} = -\frac{dx''}{dy''};$$

hence, the equation of the normal is,

$$y - y'' = -\frac{dx''}{dy''}(x - x'') \quad \ldots \quad (2.)$$

By differentiating the equation of the circle, the ellipse, the parabola, and the hyperbola, finding in each differential equation the value of $-\dfrac{dx''}{dy''}$, substituting that value in Equation (2), and reducing, we shall find the equation of the normal line to each curve.

Value of tangent, sub-tangent, normal, and sub-normal.

45. Let P be any point of a curve; TP the tangent, TR the sub-tangent, PN the normal, and RN the sub-normal.

Then, in the right-angled triangle TPR,

$$PR = TR \times \tan PTR = TR \times \frac{dy}{dx};$$

hence, $\quad TR = \dfrac{PR}{\dfrac{dy}{dx}} = y\dfrac{dx}{dy} = $ Sub-tangent.

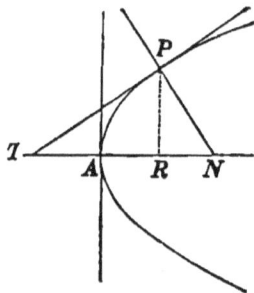

---

* Bk. I. Art. **23.**

46. The tangent TP is equal to the square root of the sum of the squares of TR and PR; hence,

$$TP = y\sqrt{1 + \frac{dx^2}{dy^2}} = \text{Tangent.}$$

47. Since TPN is a right angle, RPN is the complement of TPR; it is therefore equal to PTR, and consequently its tangent is $\frac{dy}{dx}$; hence,

$$RN = y\frac{dy}{dx} = \text{Sub-normal.}$$

48. The normal PN is equal to the square root of the sum of the squares of PR and RN; hence,

$$PN = y\sqrt{1 + \frac{dy^2}{dx^2}} = \text{Normal.}$$

49. Apply these formulas to lines of the second order, of which the general equation is,

$$y^2 = mx + nx^2.\text{*}$$

Differentiating, we have,

$$\frac{dy}{dx} = \frac{m + 2nx}{2y} = \frac{m + 2nx}{2\sqrt{mx + nx^2}};$$

substituting this value, we find,

$$TR = y\frac{dx}{dy} = \frac{2(mx + nx^2)}{m + 2nx} = \text{Sub-tangent.}$$

$$TP = y\sqrt{1 + \frac{dx^2}{dy^2}} = \sqrt{mx + nx^2 + 4\left(\frac{mx + nx^2}{m + 2nx}\right)^2}.$$

* Bk. V. Art. **42.**

$$RN = y\frac{dy}{dx} = \frac{m + 2nx}{2} = \text{Sub-normal.}$$

$$PN = y\sqrt{1 + \frac{dy^2}{dx^2}} = \sqrt{mx + nx^2 + \frac{1}{4}(m + 2nx)^2}.$$

By attributing proper values to m and n, the above formulas will become applicable to each of the conic sections. In the case of the parabola, $n = 0$, and we have,

$$TR = 2x, \qquad TP = \sqrt{mx + 4x^2},$$

$$RN = \frac{m}{2}, \qquad PN = \sqrt{mx + \frac{1}{4}m^2}.$$

Asymptotes.

50. An asymptote of a curve is a line which continually approaches the curve, and becomes tangent to it at an infinite distance from the origin of co-ordinates.

Let AX and AY be the co-ordinate axes, and

$$y - y'' = \frac{dy''}{dx''}(x - x''),$$

the equation of any tangent line, as TP.

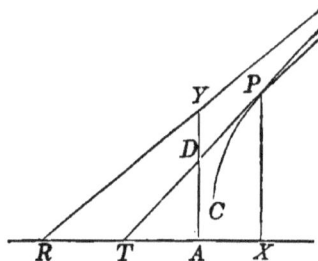

If, in the equation of the tangent, we make, in succession, $y = 0$, $x = 0$, we shall find,

$$x = AT = x'' - y''\frac{dx''}{dy''}, \qquad y = AD = y'' - x''\frac{dy''}{dx''}.$$

If the curve CPB has an asymptote RE, it is plain that the tangent PT will approach the asymptote RE, when the point of contact P, is moved along the curve from the origin of co-ordinates, and T and D will also approach the points R and Y, and will coincide with them when the co-ordinates of the point of tangency are infinite.

In order, therefore, to determine if a curve have asymptotes, we substitute in the values of AT and AD, the co-ordinates of the point which is at an infinite distance from the origin of co-ordinates. If either of the distances AT, AD, becomes finite, the curve will have an asymptote.

If both the values are finite, the asymptote will be inclined to both the co-ordinate axes; if one of the distances becomes finite and the other infinite, the asymptote will be parallel to one of the co-ordinate axes; and if they both become 0, the asymptote will pass through the origin of co-ordinates. In the last case, we shall know but one point of the asymptote, but its direction may be determined by finding the value of $\dfrac{dy}{dx}$, under the supposition that the co-ordinates are infinite.

51. Let us now examine the equation,

$$y^2 = mx + nx^2,$$

of lines of the second order, and see if these lines have asymptotes. We find,

$$AT = x - \frac{2y^2}{m + 2nx} = \frac{-mx}{m + 2nx},$$

$$AD = y - \frac{mx. + 2nx^2}{2y} = \frac{mx}{2\sqrt{mx + nx^2}},$$

which may be put under the forms,

$$AT = \frac{-m}{\dfrac{m}{x} + 2n}, \qquad\qquad AD = \frac{m}{2\sqrt{\dfrac{m}{x} + n}},$$

and making $x = \infty$, we have,

$$AR = -\frac{m}{2n}, \qquad \text{and} \qquad AE = \frac{m}{2\sqrt{n}},$$

If now we make $n = 0$, the curve becomes a parabola, and both the limits, AR, AE, become infinite; hence, the parabola has no rectilinear asymptote.

If we make n negative, the curve becomes an ellipse, and AE becomes imaginary; hence, the ellipse has no asymptote.

But if we make n positive, the equation becomes that of the hyperbola, and both the values, AR, AE, become finite. If we substitute for m its value, $\dfrac{2B^2}{A}$, and for n its value $\dfrac{B^2}{A^2}$, we shall have,

$$AR = -A, \qquad \text{and} \qquad AE = \pm B.$$

Hence, *of the lines of the second order, the hyperbola alone has asymptotes.*

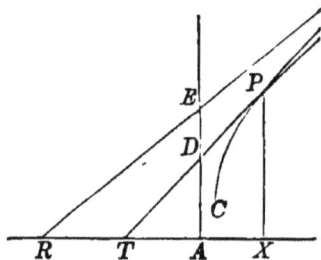

Differential of an arc.

52. We have seen that, when the points which limit any arc of a curve become consecutive, the chord, the arc, and tangent become equal (Art. **43**); therefore, *the differential of an arc is the hypothenuse* of a right-angled triangle of which the base is *dx*, and the perpendicular *dy*. Hence, if we denote any arc, referred to rectangular co-ordinates, by *z*, we have,

$$dz = \sqrt{dx^2 + dy^2} \ .\ .\ (1.) \quad \text{or,} \quad z = \int \sqrt{dx^2 + dy^2} \ .\ .\ (2.)$$

Rectification of a plane curve.

53. The *rectification* of a curve is the operation of finding its length; and when its length can be exactly expressed in terms of a linear unit, the curve is said to be *rectifiable*. To rectify a curve, given by its equation:

Differentiate the equation of the curve and find the value of dy^2 *in terms of* x *and* dx*; or of* dx^2 *in terms of* y *and* dy*, and substitute the value so found in the differential Equation* (2). *The second member will then contain but one variable and its differential; the integral will express the length of the arc in terms of that variable.*

1. Find the length of the arc of a circle in terms of the radius. The equation of a circle whose radius is 1, referred to rectangular axes, when the origin is at the centre, is,

$$x^2 + y^2 = 1.$$

Denoting the arc by z, we have,

$$dz = \sqrt{dx^2 + dy^2}, \quad \text{or,} \quad z = \int \sqrt{dx^2 + dy^2}.$$

From the equation of the circle, we have,

$$xdx + ydy = 0; \quad \text{hence,} \quad dy^2 = \frac{x^2 dx^2}{1 - x^2};$$

$$z = \int \sqrt{dx^2 + \frac{x^2 dx^2}{1 - x^2}} = \int \frac{dx}{\sqrt{1 - x^2}} = \int (1 - x^2)^{-\frac{1}{2}} dx.$$

Developing the binomial factor into a series, by the binomial formula,* multiplying by dx, and integrating, we have (Art. **42**),

$$z = \int (1-x^2)^{-\frac{1}{2}} dx = x + \frac{1x^3}{2.3} + \frac{1.3x^5}{2.4.5} + \frac{1.3.5x^7}{2.4.6.7} + \&c. + C.$$

If we suppose the origin of the integral to be at E, the corresponding value of x will be zero, and $C = 0$. If now we integrate between the limits $x = 0$, and $x = \frac{1}{2}$, we shall obtain the value of the corresponding arc in terms of the radius 1.

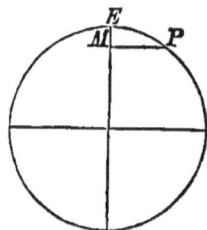

But x, or PM, is the sine of the arc EP, denoted by z; and when $x = \frac{1}{2}$, $z = 30°$; hence,

$$30° = \int_0^{\frac{1}{2}} (1 - x^2)^{-\frac{1}{2}} dx = \frac{1}{2} + \frac{1}{2.3.2^3} + \frac{1.3}{2.4.5.2^5} + \&c.,$$

---

* Bourdon, Art. **135**. University, Art. **104**.

hence,

$$\pi = 30° \times 6 = 6\left(\frac{1}{2} + \frac{1.1.1}{2.3.2^3} + \frac{1.3.1.1}{2.4.5.2^5} + \frac{1.3.5.1.1}{2.4.6.7.2^7} + \&c.\right),$$

and by taking the first ten terms of the series, we find,

$$\pi = 3.1415926. . . ,$$

a result true to the last decimal figure.

We have thus found the semi-circumference of a circle whose radius is 1, or the circumference of a circle whose diameter is 1.

2. Find the length of the arc of a parabola, whose equation is,

$$y^2 = 2px.$$

Differentiating and dividing by 2, we have,

$$ydy = pdx,$$

and consequently,

$$dx^2 = \frac{y^2}{p^2} dy^2;$$

substituting this value in the differential of the arc, we have,

$$dz = \sqrt{dy^2 + \frac{y^2}{p^2} dy^2}$$

$$= \frac{1}{p} dy \sqrt{p^2 + y^2};$$

developing the radical quantity by the binomial formula, and integrating the terms separately, we have,

$$z = \left(y + \frac{1}{2}\cdot\frac{1}{3}\frac{y^3}{p^2} - \frac{1}{2}\cdot\frac{1}{2}\cdot\frac{1}{2}\cdot\frac{1}{5}\frac{y^5}{p^4} + \frac{1}{2}\cdot\frac{1}{2}\cdot\frac{3}{2}\cdot\frac{1}{2}\cdot\frac{1}{3}\cdot\frac{1}{7}\frac{y^7}{p^6} - \&c.\right) + C.$$

If we estimate the arc from the principal vertex, z and y will be zero together, and C will be zero. If we make $y = p$, z will denote the length of the arc from the vertex to the extremity of the ordinate passing through the focus.

QUADRATURES.

54. QUADRATURE is the operation of finding the area or measure of a surface. When this measure can be found in exact terms of the unit of measure, the surface is said to be *quadrable*.

Quadrature of plane figures.

55. A *plane figure* is a portion of a plane, bounded by lines, either straight or curved.

Let O be the origin of a system of rectangular co-ordinates, and *oacdeb* any line whose equation is of the form,

$$y = f(x) \quad . \quad . \quad (1.)$$

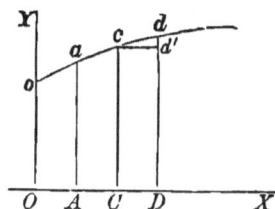

If the ordinate Oo, denoted by y, move parallel to itself, along OD as a directrix, and so change its value as always to satisfy Equation (1), it will generate the plane surface $OoacdD$, and its upper extremity will generate the line *oacd*. The *element*, or *differential* of this surface will be any one of the trapezoids, as $CcdD$, when the ordinates Cc and Dd are consecutive. If we denote the surface on the left of the ordinate Cc, by s, ds will denote the area of the trapezoid. This trapezoid is composed of the rectangle Cd', and the triangle $cd'd$; that is,

$$ds = y dx + \frac{dy dx}{2}.$$

But since the product ydx is an infinitely small quantity of the first order, and $dydx$ an infinitely small quantity of the second order, the latter may be omitted without error (Art. **20**); hence,

$$ds = ydx; \text{ that is,}$$

The differential of a plane surface is equal to the or·dinate into the differential of the abscissa.

To apply the principle enunciated in the last equation, in finding the measure of any particular plane surface:

Find the value of y in terms of x, from the equation of the bounding line; substitute this value in the differential equation, and then integrate between the required limits of x.

Nature of the Integral.

56. To comprehend the true nature of an integral, we must examine the differential from which it was derived. The differential of a plane surface is,

$$ds = ydx.$$

If we integrate between the limits $x = 0$, and $x = OD = a$, we write,

$$\int_0^a ds = \int ydx = OoacdD;$$

that is, the first member of the equation denotes the sum of all the infinitely small rectangles between the limits $x = 0$, and $x = a$; the second member,

$$\int ydx,$$

is the same thing under another form; viz.: it shows that

every value of y, between the limits $y = Oo$, and $y = Bb$, is multiplied, in succession, into each base denoted by dx; the sum of these products, each of which is ydx, is obviously the required area.

1. Perhaps the relation between the differential and the integral, may be more obvious, by observing the figure, in which a portion of it is divided into three parts, having equal bases. If we bisect each base and draw parallel ordinates, we shall have six parts; if we bisect again and draw parallel ordinates, we shall have twelve parts; if again, twenty-four; and so on.

Now, there is no difficulty in seeing that each bisection doubles the number of parts, and diminishes the value of each part; and that the sum of the parts will be constantly equal to the given area. When, therefore, each part becomes *infinitely small*, any *finite* number of them is 0; but an *infinite* number is equal to a *finite* quantity, viz.: to the given area.

Area of a rectangle.

57. Let O be the origin of a system of rectangular co-ordinates. On the axis of Y, take any distance OB equal to h. Suppose the line h to move parallel to itself, along the axis of X, as a *directrix*, until it reaches the position AC. During its motion, it will generate the rectangle OC; the foot of the line will pass over every point in the line OA, and the line itself will occupy every part of the rectangle OC.

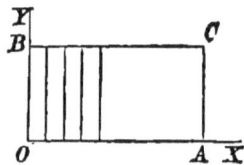

Since the equation of the line BC is,

$$y = h,$$

we shall have, for the differential of the surface,

$$ds = hdx.$$

Integrating between the limits $x = 0$, and $x = b$, **and** observing that $C = 0$, when $x = 0$, we have,

$$\int_0^b ds = \int hdx = hx = hb; \text{ that is,}$$

The area of a rectangle is equal to the product of its base by its altitude.

Area of a triangle.

58. Let ABC be a right-angled triangle, and C the origin of co-ordinates. Denote the base AB by b, and the altitude CB by h. Denote any line parallel to the base by y, and the corresponding altitude by x.

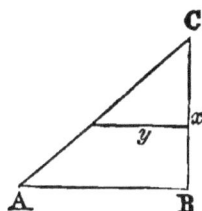

If we suppose the base AB to be moved towards the vertex of the triangle, along CB as a directrix, and so to change its value, that,

$$b : h :: y : x, \quad \text{or,} \quad y = \frac{bx}{h},$$

it is plain that it will generate the surface of the triangle. If we denote the surface by s, we have,

$$ds = ydx;$$

substituting for y its value, and integrating between the limits $x = 0$, and $x = h$, we have,

$$\int_0^h ds = \frac{b}{h}\int x\,dx = \frac{b}{h}\frac{x^2}{2} = \frac{bh}{2};$$

that is, *The area of a triangle is equal to half the product of the base by the altitude.*

Area of the parabola.

59. Find the area of any portion of the common parabola whose equation is,

$$y^2 = 2px; \quad \text{whence,} \quad y = \sqrt{2px}.$$

This value of y being substituted in the differential equation (Art. **55**), gives (Art. **36**),

$$\int ds = \int \sqrt{2px}\,dx = \sqrt{2p}\int x^{\frac{1}{2}}\,dx = \frac{2\sqrt{2p}}{3}x^{\frac{3}{2}} + C;$$

or, $\quad s = \dfrac{2\sqrt{2px}\times x}{3} \qquad = \dfrac{2}{3}xy + C.$

If we estimate the area from the principal vertex, where $x = 0$, and $y = 0$, we have, $C = 0$, and denoting the particular integral by s', we shall have,

$$s' = \frac{2}{3}xy; \quad \text{that is,}$$

The area of any portion of the parabola, estimated from the vertex, is equal to $\dfrac{2}{3}$ of the rectangle of the abscissa and ordinate of the extreme point. The curve is, therefore, QUADRABLE.

1. To find the area of a parabola from the vertex to the double ordinate through the focus. We have, for these limits, $x = 0$ and $x = \frac{1}{2}p$. Denoting the integral by s'',

we have,
$$\int_0^{\frac{1}{2}p} ds = s'' = \frac{1}{3}p^2,$$

which denotes the area bounded by the curve, the axis, and the ordinate; hence, if we double it, we shall have the required area; or,

$$2s'' = \frac{2}{3}p^2 = \frac{4}{6}p^2 = \frac{1}{6}(2p)^2;$$

That is, *The area is equal to one-sixth of the square described on the parameter of the axis.*

2. If the area be estimated from the ordinate through the focus, where $x = \frac{1}{2}p$, and $y = p$, C must have such a value as to reduce the first member to 0: for, this is the origin of the integral.

We have,
$$\int ds = \frac{2}{3}xy + C;$$

and for the particular case of the focus,

$$\int ds = \frac{2}{3} \times \frac{1}{2}p \times p + C = \frac{1}{3}p^2 + C; \text{ hence,}$$

$$\frac{1}{3}p^2 + C = 0; \quad \text{or,} \quad C = -\frac{1}{3}p^2.$$

Hence, the integral from $x = \frac{1}{2}p$ to any value of x is,

$$\int_{\frac{1}{2}p}^{x} ds = \frac{2}{3}xy - \frac{1}{3}p^2.$$

Area of the circle.

60. The equation of the circle referred to its centre and rectangular axes is,

$$y^2 = r^2 - x^2; \quad \text{or,} \quad y = \sqrt{r^2 - x^2};$$

hence, the differential equation of the area (Art. **57**) is,

$$ds = \left(\sqrt{r^2 - x^2}\right)dx \quad \ldots \ldots \text{(1.)}$$

in which the origin of the area is at the secondary dia-meter, where $x = 0$.

From Formula $\mathcal{B}_3$ page 189, we have,

$$\int \left(\sqrt{r^2 - x^2}\right)dx = \frac{1}{2}x(r^2 - x^2)^{\frac{1}{2}} + \frac{1}{2}r^2 \int (r^2 - x^2)^{-\frac{1}{2}}dx.$$

But, by Formula (13), Art. **99**, we have,

$$\int (r^2 - x)^{-\frac{1}{2}}dx = \int \frac{dx}{\sqrt{r^2 - x^2}} = \sin^{-1}\frac{x}{r} + C;$$

whence, by substitution, we have,

$$s = \frac{1}{2}x(r^2 - x^2)^{\frac{1}{2}} + \frac{1}{2}r^2 \sin^{-1}\frac{x}{r} + C. \quad \text{(2.)}$$

Estimating the area from the secondary diameter, where $x = 0$, we have, $C = 0$.

If we integrate between the limits of $x = 0$, and $x = r$, we shall have one quarter of the area of the circle. When we make $x = r$, in Equation (2), the first term in the second member becomes 0; and in the second term, $\frac{x}{r}$ becomes 1, and the arc whose sine is 1, is 90°, which is denoted by $\frac{\pi}{2}$, to the radius 1; hence,

$$\int_0^r ds = \frac{1}{2}r^2 \sin^{-1} 1 = \frac{1}{2}r^2 \times \frac{\pi}{2}; \text{ or,}$$

$$\text{Area of the circle } = 4\left(\frac{1}{2}r^2 \times \frac{\pi}{2}\right) = r^2\pi.$$

Area of the ellipse.

61. The equation of the ellipse, referred to its centre and axes is,

$$A^2y^2 + B^2x^2 = A^2B^2 \; ; \quad \text{hence,}$$

$$y = \frac{B}{A}\sqrt{A^2 - x^2},$$

and the differential equation of the area is,

$$ds = \frac{B}{A}(A^2 - x^2)^{\frac{1}{2}}dx.$$

The second member of this equation differs from the second member of Equation (1), of the last Article, only in the constant coefficient $\dfrac{B}{A}$, and the constant A^2 for r^2, within the parenthesis; hence, the integral of that expression becomes the integral of this, by multiplying it by $\dfrac{B}{A}$, and changing r into A; that is,

$$\int ds = \frac{B}{A}\left(\frac{1}{2}A^2 \times \frac{\pi}{2}\right) = \frac{AB\pi}{4}\; ; \quad \text{hence,}$$

$$\text{Area of ellipse} = \frac{4AB\pi}{4} = A.B.\pi\; ; \quad \text{that is,}$$

The area of an ellipse is equal to the product of its semi-axes multiplied by π.

1. Let Q denote the area of a circle described on the transverse axis, and Q' the area of a circle described on the conjugate axis; then,

$$A^2\pi = Q, \quad \text{and} \quad B^2\pi = Q'\; ; \quad \text{hence,}$$

$$A^2 B^2 \pi^2 = Q Q', \quad \text{and} \quad A B \pi = \sqrt{Q \times Q'}; \text{ that is,}$$

The area of an ellipse is a mean proportional between the two circles described on its axes.

QUADRATURE OF SURFACES OF REVOLUTION.

62. Let *oacdeb* be a plane curve, *OB* the axis of abscissas, and *Oo*, *Aa*, *Cc*, &c., consecutive ordinates; then, *oa*, *ac*, *cd*, &c., will be elementary arcs. The surface described by either of these arcs, while the curve revolves around the axis *OB*, will be an element of the surface. We have seen, that when the ordinates are consecutive, the chord, the arc, and the tangent, are equal (Art. **43**); hence, the surface described by any arc, as *ac*, is equal to that described by the chord; that is, equal to the surface of the frustum of a cone, the radii of whose bases are $Aa = y$, $Cc = y + dy$, and of which the slant height $ac = \sqrt{dx^2 + dy^2}$. Hence, if we denote the surface by s, we have,*

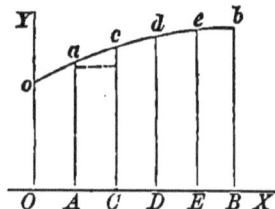

$$ds = \pi(2y + 2y + 2dy) \times \tfrac{1}{2}\sqrt{dx^2 + dy^2};$$

or, omitting $2dy$ (Art. **20**),

$$ds = 2\pi y \sqrt{dx^2 + dy^2}; \text{ that is,}$$

The differential of a surface of revolution is equal to the circumference of a circle perpendicular to the axis, into the differential of the arc of the meridian curve.

---

* Leg., Bk. VIII. P. 4.

Therefore, to find the measure of any surface of revolution:

Find the values of y and dy, from the equation of the meridian curve, in terms of x and dx; then substitute these values in the differential equation, and integrate between the proper limits of x.

Surface of a cylinder.

63. If the rectangle AC be revolved around the side AB, DC will generate the surface of a cylinder.

Since the generatrix is parallel to the axis AB, its equation will be,

$$y = b, \quad \text{and hence,} \quad dy = 0.$$

Substituting these values in the differential equation of the surface, we have,

$$\int ds = \int 2\pi y \sqrt{dx^2 + dy^2} = \int 2\pi b dx = 2\pi bx + C.$$

If we suppose A to be the origin of co-ordinates, $C = 0$, and integrating between the limits $x = 0$ and $x = h$, we have,

$$s = 2b\pi h;$$

that is, *The measure of the surface of a cylinder is equal to the circumference of its base into the altitude.*

Surface of the cone.

64. If the right-angled triangle CBA be revolved around the axis AC, CB will generate the convex surface of a cone.

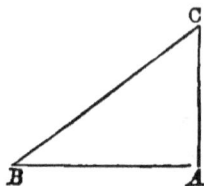

If we suppose C to be the origin of co-ordinates, the equation of BC will be,

$$y = ax, \quad \text{and} \quad dy = adx.$$

Substituting these values in the differential equation of the surface, we have,

$$\int ds = \int 2\pi ax\sqrt{dx^2 + a^2 dx^2} = \int 2\pi axdx\sqrt{1 + a^2} + C,$$

$$\text{(Art. 35)} \quad . \; . \; . \; . \; . \; . = \pi ax^2\sqrt{1 + a^2} + C.$$

Estimating the surface from the vertex, where $x = 0$, we have, $C = 0$, and

$$s = \pi ax^2\sqrt{1 + a^2}.$$

If we make $x = h = AC$, and $BA = b$, we have, $a = \dfrac{b}{h}$, and consequently,

$$s = \pi\frac{b}{h}\, h^2\sqrt{1 + \frac{b^2}{h^2}} = \frac{2\pi b\sqrt{h^2 + b^2}}{2} = 2\pi b \times \frac{BC}{2}.$$

that is, *The convex surface of a cone is equal to the circumference of the base into half the slant height.*

Surface of the sphere.

65. To find the surface of a sphere. The equation of the meridian curve, referred to the centre, is,

$$x^2 + y^2 = R^2.$$

By differentiating, we have,

$$x dx + y dy = 0;$$

hence,

$$dy = - \frac{x dx}{y}, \quad \text{and} \quad dy^2 = \frac{x^2 dx^2}{y^2}.$$

Substituting for dy^2 its value, in the differential of the surface, which is,

$$ds = 2\pi y \sqrt{dx^2 + dy^2},$$

we have,

$$\int ds = \int 2\pi y \sqrt{dx^2 + \frac{x^2}{y^2} dx^2} = \int 2\pi R dx = 2\pi R x + C.$$

If we estimate the surface from the plane passing through the centre, and perpendicular to the axis of X, we shall have,

$$s = 0, \quad \text{for} \quad x = 0, \quad \text{and consequently,} \quad C = 0.$$

To find the entire surface of the sphere, we must integrate between the limits $x = + R$, and $x = - R$, and then take the sum of the integrals, without reference to their algebraic signs; for, these signs only indicate the position of the parts of the surface with respect to the plane passing through the centre.

Integrating between the limits,

$$x = 0, \quad \text{and} \quad x = + R,$$

we find, $s = 2\pi R^2$;

and integrating between the limits $x = 0$, and $x = -R$, there results,

$$s = -2\pi R^2;$$

hence,

$$\text{Surface} = 4\pi R^2 = 2\pi R \times 2R;$$

that is, *Equal to four great circles, or equal to the curved surface of the circumscribing cylinder.*

1. The two equal integrals,

$$s = 2\pi R^2, \quad \text{and} \quad s = -2\pi R^2,$$

indicate that the surface is divided into two equal parts by the plane passing through the centre.

Surface of the paraboloid.

66. To find the surface of the paraboloid of revolution. Take the equation of the meridian curve,

$$y^2 = 2px,$$

which being differentiated, gives,

$$dx = \frac{ydy}{p}, \quad \text{and} \quad dx^2 = \frac{y^2dy^2}{p^2}.$$

Substituting this value of dx in the differential of the surface, (Art. **62**), we have,

$$ds = 2\pi y \sqrt{\left(\frac{y^2 + p^2}{p^2}\right)}dy = \frac{2\pi}{p}ydy\sqrt{y^2 + p^2}.$$

But we have found (Art. **41**),

$$\int \frac{2\pi}{p} y\,dy \sqrt{y^2 + p^2} = \frac{2\pi}{3p}(y^2 + p^2)^{\frac{3}{2}} + C;$$

hence,

$$s = \frac{2\pi}{3p}(y^2 + p^2)^{\frac{3}{2}} + C.$$

If we estimate the surface from the vertex, at which point $y = 0$, we shall have,

$$0 = \frac{2\pi p^2}{3} + C, \quad \text{whence,} \quad C = -\frac{2\pi p^2}{3};$$

and integrating between the limits,

$$y = 0, \quad \text{and} \quad y = b,$$

we have,

$$s = \frac{2\pi}{3p}[(b^2 + p^2)^{\frac{3}{2}} - p^3].$$

Surface of the ellipsoid.

67. To find the surface of an ellipsoid described by revolving an ellipse about the transverse axis.

The equation of the meridian curve is,

$$A^2y^2 + B^2x^2 = A^2B^2,$$

whence,

$$dy = -\frac{B^2}{A^2}\frac{x\,dx}{y} = -\frac{B}{A}\frac{x\,dx}{\sqrt{A^2 - x^2}};$$

substituting the square of dy in the differential of the surface, and for y its value,

$$\frac{B}{A}\sqrt{A^2 - x^2},$$

we have,

$$ds = 2\pi\frac{B}{A^2}dx\sqrt{A^4 - (A^2 - B^2)x^2}; \quad (1.)$$

hence, $\int ds = 2\pi\dfrac{B}{A^2}\sqrt{A^2 - B^2}\int dx\sqrt{\dfrac{A^4}{A^2 - B^2} - x^2}.$

Put, $2\pi\dfrac{B}{A^2}\sqrt{A^2 - B^2} = D,$ a constant quantity;

and $\dfrac{A^4}{A^2 - B^2} = R^2,$ also a constant,

and we have,

$$\int ds = D\int dx\sqrt{R^2 - x^2}.$$

With C, the centre of the meridian curve, and the radius R, describe a semi-circle. Then, $\int dx\sqrt{R^2 - x^2}$, is a circular segment of which the abscissa is x, and radius R.

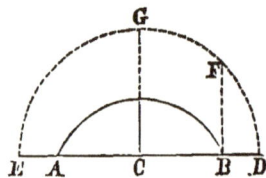

If, then, we estimate the surface of the ellipsoid from the plane passing through the centre, and estimate the area of the circular segment from the same plane, any portion of the surface of the ellipsoid will be equal to the corresponding portion of the circle, multiplied by the constant D. Hence, if we integrate the expression,

$$\int dx \sqrt{R^2 - x^2},$$

between the limits $x = 0$, and $x = A$, we shall have the area of the segment $CGFB$, which denote by D'. Hence,

$\frac{1}{2}$ surface ellipsoid $= D \times D'$; and

Surface $\qquad = 2D \times D'$.

1. If we make $A = B$, in Equation (1), the ellipsoid becomes a sphere, and we have,

$$s = \int 2\pi R dx = 2\pi Rx + C.$$

If we estimate the surface from the plane passing through the centre, $C = 0$, and integrate between the limits $x = 0$, and $x = R$, we have,

$\frac{1}{2}$ surface of sphere $= 2\pi R^2$; hence,

Surface $\qquad = 4\pi R^2$.

CUBATURE OF VOLUMES OF REVOLUTION.

68. CUBATURE is the operation of finding the measure of a volume. When this measure can be found in exact terms of the measuring cube, the volume is said to be *cubable*.

69. A *volume of revolution* is a volume generated by the revolution of a plane figure about a fixed line, called the *axis*.

If the plane figure $OoacdebB$, be revolved about the axis of X, it will generate a volume of revolution.

Let us suppose the ordinates Aa, Cc, Dd, &c., to be consecutive. During the revolution, any element of the surface, as, $AacC$, will generate the frustum of a cone, of which the radii of the bases are $Aa = y$, $Cc = y + dy$, and the altitude, $AC = dx$. This frustum will be an element of the volume, and will have for its measure,*

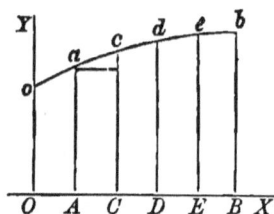

$$\frac{\pi}{3}[y^2 + (y + dy)^2 + y(y + dy)]dx.$$

If we denote the volume by V, develop the terms within the parenthesis, multiply by dx, and then reject all the terms containing the infinitely small quantities of the second order (Art. **20**), we shall have,

$$dV = \pi y^2 dx.$$

The area of a circle described by any ordinate y, is πy^2†; hence, *The differential of a volume of revolution is equal to the area of a circle perpendicular to the axis into the differential of the axis.*

The differential of a volume generated by the revolution of a plane figure about the axis of Y, is $\pi x^2 dy$.

70. To find the value of V for any given volume:

Find the value of y^2 in terms of x, from the equation of the meridian curve; substitute this value in the differential equation, and then integrate between the required limits of x.

---

* Leg., Bk. VIII. P. 6.　　† Leg., Bk. V. Prop. 16.

EXAMPLES.

1. Find the volume of a right cylinder with a circular base, whose altitude is h and the radius of whose base is r.

We have for the differential of the volume,

$$dV = \pi y^2 dx ;$$

and since $y = r$, we have,

$$\int dV = \int \pi r^2 dx ;$$

integrating between the limits $x = 0$, and $x = h$,

$$\int_0^h dV = V = \pi r^2 x = \pi r^2 h ; \text{ that is,}$$

*The measure of the volume of a cylinder is equal to the area of its base multiplied by the altitude.**

2. Find the volume of a right cone with a circular base, whose altitude is h, and the radius of the base, r.

If we suppose the vertex of the cone to be at the origin of co-ordinates, and the axis to coincide with the axis of abscissas, we shall have,

$$y = ax, \quad \text{or,} \quad y = \frac{r}{h}x, \quad \text{and} \quad y^2 = \frac{r^2}{h^2}x^2 ;$$

substituting this value of y^2, we have,

$$\int dV = \int \pi \frac{r^2}{h^2}x^2 dx.$$

* Legendre, Bk. VIII. Prop. 2.

Integrating between the values $x = 0,$ and $x = h,$

$$\int_0^h dV = V = \pi\frac{r^2}{h^2}\frac{x^3}{3} = \pi r^2 \times \frac{h}{3}; \text{ that is,}$$

The measure of the volume of a cone is equal to the area of the base into one-third of the altitude.

3. To find the volume of a prolate spheroid. †

The equation of the meridian curve is,

$$A^2y^2 + B^2x^2 = A^2B^2; \quad \text{hence,} \quad y^2 = \frac{B^2}{A^2}(A^2 - x^2).$$

and $\qquad dV = \pi\dfrac{B^2}{A^2}(A^2 - x^2)dx;$ hence,

$$V = \frac{\pi B^2}{A^2}\Big(A^2x - \frac{x^3}{3}\Big) + C,$$

$$= \frac{\pi B^2}{3A^2}(3A^2x - x^3) + C.$$

If we estimate the volume from the plane passing through the centre, we have, for $x = 0,$ $V = 0,$ and consequently, $C = 0;$ and taking the integral between the limits $x = 0,$ and $x = A,$ we have,

$$\int_c^A dV = \frac{2}{3}\pi B^2 \times A;$$

which is half the volume; consequently, the entire volume,

$$2V = \frac{2}{3}\pi B^2 \times 2A$$

---

* Legendre, Bk. VIII. Prop. 6.   † Bk. VI. Art. **37.**

But, πB^2 expresses the area of a circle described on the conjugate axis, and $2A$ is the transverse axis; hence,

The volume of a prolate spheroid is equal to two-thirds of the circumscribing cylinder.

1. If an ellipse be revolved around the conjugate axis, it will describe an oblate spheroid, and we shall have,

$$\int dV = \int \pi x^2 dy;$$

substituting for x^2, and integrating, we have,

$$2V = \frac{2}{3}\pi A^2 \times 2B;$$

that is, two-thirds of the circumscribing cylinder.

2. If we compare the two results together, we find,

oblate spheroid : prolate spheroid :: $A : B$.

3. If we make $B = A$, the ellipsoid becomes a sphere whose diameter is the transverse axis. Then,

$$2V = \frac{2}{3}\pi R^2 \times D = \frac{1}{6}\pi D^3;$$

that is, *Equal to two-thirds of the circumscribing cylinder, or to one-sixth of π into the cube of the diameter.*

4. Find the volume of a paraboloid. The equation of the meridian curve is,

$$y^2 = 2px; \text{ hence,}$$

$$dV = 2\pi px dx, \quad \text{and} \quad V = \pi px^2.$$

If we estimate the volume from the vertex, $C = 0$. If we integrate between the limits $x = 0$, and $x = h$, and designate by b, the ordinate corresponding to the abscissa $x = h$, we have,

$$V = \pi p h^2 = \pi b^2 \times \frac{h}{2}; \quad \text{that is,}$$

equal to half the cylinder having the same base and altitude.

Prism and Pyramid.

1. Let $ABCDE$ be any polygon, and FH a line perpendicular to the plane of the base. If the polygon move along the line FH, parallel to itself, it will generate a prism. If we denote the volume by V, the area of the base by b, and the indefinite line HF by x, we shall have,

$$dV = b dx.$$

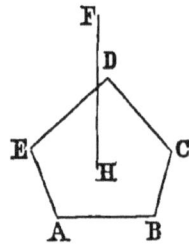

and, integrating between the limits $x = 0$, and $x = h$,

$$\int_0^h dV = \int b dx = b \times x = b \times h.$$

2. If we suppose the base so to vary, as it moves along the line FH, as to bear a constant ratio to the square of its distance from the point F, it will generate the volume of a pyramid, of which F is the vertex and $ABCDE$ the base.* If we denote the variable generatrix, at any point, by y, and its distance from the vertex by x, we have,

$$dV = y dx.$$

But, $b : y :: h^2 : x^2$; hence, $y = \dfrac{b}{h^2} \times x^2$;

therefore, $$dV = \frac{b}{h^2} \times x^2 dx;$$

and integrating between the limits $x = 0$, and $x = h$, we have,

$$\int_0^h dV = \frac{b}{h^2} \int x^2 dx = \frac{b}{h^2} \times \frac{x^3}{3} = b \times \frac{h}{3}.$$

---

* Legendre, Bk. VII. P. 3. Cor. 1.

SECTION IV.

Successive Differentials.

71. If u denotes any function, and x the independent variable, we have seen that the differential coefficient P, is, in general, a function of x (Art. **23**). It may therefore be differentiated, and a new differential coefficient will thus be obtained, which is called the *second differential coefficient.*

In passing from the function u to the first differential coefficient, the exponent of x is diminished by 1, in every term where x enters (Art. **30**); hence, the *relation* between the primitive function u and the variable x, is different from that which exists between the first differential coefficient and x. Hence, the same change in x, will occasion *different degrees* of change in the *primitive function* and in the *first differential coefficient.*

The second differential coefficient will, in general, be a function of x, exhibiting a still different relation; hence, a new differential coefficient may be formed from it, which may also be a function of x; and so on, for succeeding differential coefficients.

If we designate the successive differential coefficients by

$$p, \quad q, \quad r, \quad s, \quad \&c.,$$

we shall have,

$$\frac{du}{dx} = p, \qquad \frac{dp}{dx} = q, \qquad \frac{dq}{dx} = r, \ \&c. \ ; \ \text{ and}$$

$$du = pdx, \qquad dp = qdx, \qquad dq = rdx.$$

But the differential of p may be obtained by differentiating its value $\frac{du}{dx}$, regarding the denominator dx as constant; we therefore have,

$$d\left(\frac{du}{dx}\right) = dp, \qquad \text{or,} \qquad \frac{d^2u}{dx} = dp \ ;$$

substituting for dp its value, and dividing by dx,

$$\frac{d^2u}{dx^2} = q.$$

The notation, d^2u, indicates that the function u has been differentiated twice; it is read, *second differential of u.* The denominator dx^2, denotes *the square of the differential of x,* and not the differential of x^2. It is read: differential of x, *squared.*

If we differentiate the value of q, we have,

$$d\left(\frac{d^2u}{dx^2}\right) = dq, \qquad \text{or,} \qquad \frac{d^3u}{dx^2} = \boldsymbol{dq} \ ;$$

hence,
$$\frac{d^3u}{dx^3} = r, \ \&c. \ ;$$

and in the same manner we may find,

$$\frac{d^4u}{dx^4} = s.$$

The third differential coefficient, $\frac{d^3u}{dx^3}$, is read: third differential of u, divided by dx cubed; and the differential coefficients which succeed it are read in a similar manner.

Hence, the successive differential coefficients are,

$$\frac{du}{dx} = p, \qquad \frac{d^2u}{dx^2} = q, \qquad \frac{d^3u}{dx^3} = r, \qquad \frac{d^4u}{dx^4} = s, \quad \&c.,$$

from which we see, that each differential coefficient is derived from the one that immediately precedes it, in the same way as the first is derived from the primitive function.

The differentials of the different orders are obtained by multiplying the differential coefficients by the corresponding powers of dx; thus,

$$\frac{du}{dx}\,dx = \text{1st differential of } u,$$

$$\frac{d^2u}{dx^2}\,dx^2 = \text{2d differential of } u,$$

$$\cdot \quad \cdot \quad \cdot \quad \cdot \quad \cdot \quad \cdot \quad \cdot \quad \cdot \quad \cdot \quad \cdot$$

$$\frac{d^nu}{dx^n}\,dx^n = n\text{th differential of } u.$$

1. Find the differential coefficients in the function,

$$u = ax^3.$$

$$\frac{du}{dx} = 3ax^2 = p,$$

$$\frac{d^2u}{dx^2} = 6ax = q,$$

$$\frac{d^3u}{dx^3} = 6a = r.$$

2. Find the differential coefficients in the function,

$$u = ax^n.$$

The first differential coefficient is,

$$\frac{du}{dx} = nax^{n-1}.$$

Since n, a, and dx, are constants, we have for the second differential coefficient,

$$\frac{d^2u}{dx^2} = n(n-1)ax^{n-2};$$

and for the third,

$$\frac{d^3u}{dx^3} = n(n-1)(n-2)ax^{n-3};$$

and for the fourth,

$$\frac{d^4u}{dx^4} = n(n-1)(n-2)(n-3)ax^{n-4}.$$

It is plain, that when n is a positive integral number, the function

$$u = ax^n,$$

will have n differential coefficients.   For, when n differentiations have been made, the exponent of x in the second member will be 0; hence, the nth differential coefficient will be a constant, and the succeeding ones will be 0.   Thus,

$$\frac{d^n u}{dx^n} = n(n-1)(n-2)(n-3)\ldots\ldots a.1$$

and $$\frac{d^{n+1}u}{dx^{n+1}} = 0.$$

Sign of the first differential coefficient.

72. If we have a curve whose equation is,

$$y = f(x),$$

and give to x any increment h, we have (Art. **13**),

$$\frac{y'-y}{h} = \frac{f(x+h)-f(x)}{h},$$

and passing to the consecutive values,

$$\frac{dy}{dx} = \tan \alpha.$$

If we so place the origin of co-ordinates that the curve shall lie within the first angle, h will be positive, and $y' - y$ will be positive at all points where the curve recedes from the axis of X, and negative where it approaches the axis; and this is true for consecutive as well as for other values. Hence, *the curve will recede from the axis of X when the first differential coefficient is positive, and approach the axis when that coefficient is negative.*

The general proposition for all the angles and every possible relation of y and x, is this:

The curve will recede from the axis of X when the ordinate and first differential coefficient have the same sign, and approach it when they have different signs.

1. To determine whether a given curve, as ABC, recedes from, or approaches to the axis of X, at any point, as C: Find, from the equation of the curve, the first differential coefficient, and see whether it is positive or negative.

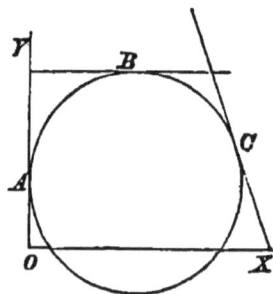

2. If the tangent becomes parallel to the axis of X at any point, as B,

$$\frac{dy}{dx} = \tan \alpha = 0; \quad \text{hence,} \quad \alpha = 0.$$

If the tangent becomes perpendicular to the axis of X, at any point, as A

$$\frac{dy}{dx} = \tan \alpha = \infty; \quad \text{hence,} \quad \alpha = 90°.$$

Sign of the second differential coefficient.

73. A curve is *convex* towards the axis of abscissas when it lies between the chord and the axis; and *concave*, when the chord lies between the curve and the axis.

1.   2.

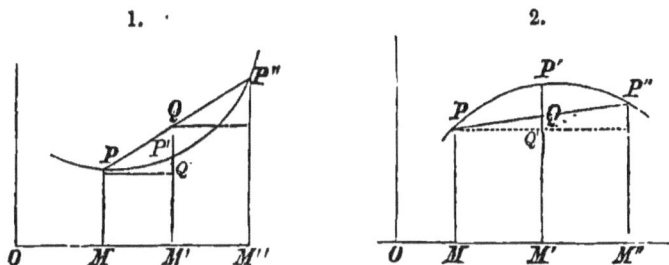

Figures (1) and (2) denote two curves, the one convex and the other concave towards the axis of X.

Let PM be any ordinate of either curve, $P'M'$ an ordinate consecutive with it, and $P''M''$ an ordinate consecutive with $P'M'$.

If we designate the ordinate PM by y, $P'Q'$ will be denoted by dy (Art. **21**), and we shall have,

$$P'M' = y + dy;$$

and since $P''M''$ is consecutive with $P'M'$,

$$P''M'' = y + \ dy + d(y + dy)$$
$$= y + 2dy + d^2y.$$

Since, $MM' = M'M'' = dx$, $QM' = \dfrac{MP + P''M''}{2}$;

hence, $QM' = \dfrac{y + y + 2dy + d^2y}{2} = y + dy + \dfrac{d^2y}{2}$,

and $QM' - P'M' = QP' = \dfrac{d^2y}{2}.$

In the case of *convexity*, $QM' > P'M'$, and then, d^2y is positive.

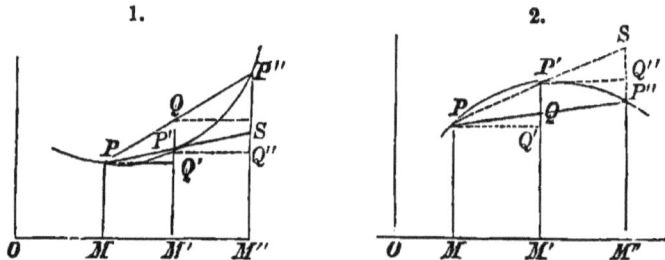

In the case of *concavity*, $QM' < P'M'$, and then, d^2y is negative; and since dx^2 is always positive, the second differential coefficient will have the same sign as the second differential of y.

If we take the case in which the ordinates are negative, the second differential coefficient will still have the same sign as the ordinate, when the curve is convex, and a different sign when it is concave. Hence,

The second differential coefficient will have the same sign as the ordinate when the curve is convex towards the axis of abscissas, and a contrary sign when it is concave.

1. The second differential of y is derived from dy in the same way that dy is derived from y (Art. **72**); viz.: by producing the chord PP', and finding the difference of the consecutive values of $P''Q''$ and SQ'', which is $P''S$.

The co-ordinates x and y determine a single point of the curve, as P; these, in connection with dx and dy, determine a second point, P', consecutive with the first; and these two sets of values, in connection with the second differential of y, determine a third point, P'', consecutive with P'.

Hence, the co-ordinates x and y, and the first and second differential coefficients, *always determine three consecutive points of a curve.*

2. When the curve is convex towards the axis of abscissas, the tangent of the angle which the tangent line makes with the axis of X, is an increasing function of x; hence, its differential coefficient, that is, the *second* differential of the function, ought to be, as we have found it, positive (Art. **19**).

When the curve is concave, the first differential coefficient is a decreasing function of the abscissas; hence, the second differential coefficient should be negative (Art. **19**).

Applications.

74. The equation of the circle, referred to its centre and rectangular axes, is,

$$x^2 + y^2 = R^2 ; \qquad \text{hence,} \qquad \frac{dy}{dx} = -\frac{x}{y}.$$

Placing $\qquad -\frac{x}{y} = 0, \qquad$ we have, $\qquad x = 0.$

Substituting this value of x in the equation of the circle, we have,

$$y = \pm R ;$$

hence, the tangent is parallel to the axis of abscissas at the two points where the axis of ordinates intersects the circumference.

If we make, $\qquad \dfrac{dy}{dx} = -\dfrac{x}{y} = \infty, \qquad$ we have, $\quad y = 0 ;$

substituting this value in the equation of the circle,

$$x = \pm R ; \quad \text{hence,}$$

the tangent is perpendicular to the axis of abscissas at the points where the axis intersects the circumference.

1. For the second differential coefficient, we find,

$$\frac{d^2 y}{dx^2} = -\frac{R^2}{y^3},$$

which will be negative when y is positive, and positive when y is negative. Hence, the circumference of the circle is concave towards the axis of abscissas.

2. If we apply the same process to the equation of the ellipse, of the parabola, and of the hyperbola, we shall find that the tangents, at the principal vertices, are parallel to the axes of ordinates; that the second differential coefficient and ordinate, in all the cases, except that of the opposite hyperbolas, have contrary signs; and hence, *all these curves, except the conjugate hyperbolas, are concave towards the axis of abscissas.*

MACLAURIN'S THEOREM.

75. MACLAURIN'S THEOREM explains the method of developing into a series any function of a single variable.

Let u denote any function of x, as, for example,

$$u = (a + x)^m \quad . \quad . \quad . \quad . \quad (1.)$$

It is required to develop this, or any other function of x, into a series of the form,

$$u = A + Bx + Cx^2 + Dx^3 + Ex^4 + \&c. \quad . \quad . \quad (2.)$$

in which A, B, C, D, &c., are independent of x, and arbitrary functions of the constants which enter into the

second member of Equation (1). When these coefficients are found, the form of the series will be known.

Since the coefficients, A, B, C, &c., are, by hypothesis,
independent of x, each will have the same value for
$x = 0$, as for any other value of x; hence, it is only
necessary to determine them for $x = 0$.

If we make $x = 0$, in Equation (2), all the terms
in the second member, after the first, will become zero,
and the second member will reduce to A, which is what
the function u becomes in Equation (1), when $x = 0$
That value is thus indicated:

$$(u)_{x=0} = A.$$

If we find the successive differential coefficients of u,
from Equation (2), we shall have,

$$\frac{du}{dx} = B + 2Cx + 3Dx^2 + 4Ex^3 + \text{&c.}$$

$$\frac{d^2u}{dx^2} = 2C + 2.3Dx + 3.4Ex^2 + \text{&c.}$$

$$\frac{d^3u}{dx^3} = 2.3D + 2.3.4Ex + \text{&c.}$$

$$\text{&c.,} \qquad \text{&c. ;}$$

whence, $\qquad A = (u)_{x=0}$

$$B = \left(\frac{du}{dx}\right)_{x=0}$$

$$C = \frac{1}{1.2}\left(\frac{d^2u}{dx^2}\right)_{x=0}$$

$$D = \frac{1}{1.2.3}\left(\frac{d^3u}{dx^3}\right)_{x=0} \qquad \text{&c.,} \qquad \text{&c. ;}$$

hence,

$$u = (u)_{x-0} + \left(\frac{du}{dx}\right)_{x-0} x + \frac{1}{1.2}\left(\frac{d^2u}{dx^2}\right)_{x-0} x^2$$

$$+ \frac{1}{1.2.3}\left(\frac{d^3u}{dx^3}\right)_{x-0} x^3 + \&c. \quad \cdots \quad (2.)$$

which is Maclaurin's Formula. In applying the formula, we omit the expressions $x = 0$, although *the coefficients are always found under this hypothesis.*

EXAMPLES.

1. Develop $(a + x)^m$, by Maclaurin's Formula,

$$A = a^m,$$

$$B = \left(\frac{du}{dx}\right)' = m(a + x)^{m-1} = ma^{m-1},$$

$$C = \frac{1}{2}\left(\frac{d^2u}{dx^2}\right) = \frac{m(m-1)}{1.2}(a + x)^{m-2} = \frac{m(m-1)}{1.2}a^{m-2},$$

$$D = \frac{1}{1.2.3}\left(\frac{d^3u}{dx^3}\right) = \frac{m}{1}\frac{(m-1)}{2}\frac{(m-2)}{3}(a + x)^{m-3}$$

$$= \frac{m}{1}\frac{(m-1)}{2}\frac{(m-2)}{3}a^{m-3},$$

$$\&c., \qquad \&c., \qquad \&c.$$

Substituting these values in Equation (2), we have,

$$(a + x)^m = a^m + ma^{m-1}x + \frac{m}{1}\frac{(m-1)}{2}a^{m-2}x^2$$

$$+ \frac{m}{1}\frac{(m-1)}{2}\frac{(m-2)}{3}a^{m-3}x^3 + \&c.;$$

the same result as found by the Binomial Formula.

2. If the function is of the form,

$$u = \frac{1}{a+x} = (a+x)^{-1} = a^{-1}\left(1 + \frac{x}{a}\right)^{-1}.$$

we find,

$$A = \frac{1}{a},$$

$$B = \left(\frac{du}{dx}\right)_{x=0} = -1(a+x)^{-2} = -\frac{1}{(a+x)^2} = -\frac{1}{a^2},$$

$$C = \frac{1}{2}\left(\frac{d^2u}{dx^2}\right)_{x=0} = \frac{-1 \times -2(a+x)^{-3}}{2} = \frac{1}{a^3},$$

$$D = \frac{1}{2.3}\left(\frac{d^3u}{dx^3}\right)_{x=0} = \frac{-1 \times -2 \times -3(a+x)^{-4}}{2.3} = -\frac{1}{a^4},$$

&c.,   &c.,   &c.

Substituting these values in Maclaurin's Formula,

$$\frac{1}{a+x} = \frac{1}{a} - \frac{x}{a^2} + \frac{x^2}{a^3} - \frac{x^3}{a^4} + \frac{x^5}{a^6} - \frac{x^7}{a^8} + \&c.$$

3. Develop into a series, the function,

$$u = \sqrt{a^2 + x^2} = a\left(1 + \frac{x^2}{a^2}\right)^{\frac{1}{2}}.$$

4. Develop into a series, the function,

$$u = \sqrt[3]{(a^2 - x^2)^2} = a^{\frac{4}{3}}\left(1 - \frac{x^2}{a^2}\right)^{\frac{2}{3}}.$$

NOTE. 76. Maclaurin's Formula has been demonstrated under the supposition, that in Equation (2) the coefficients are independent of x, and that the equation is

true for every possible value that can be attributed to
x. If, then, the function u becomes infinite, when $x = 0$,
the equation cannot be satisfied; neither can it be, if
any one of the differential coefficients becomes infinite.
Hence, any form of the function which produces either
of these results, is excluded from the formula of Mac·
laurin. The functions,

$$u = \log x, \qquad u = \cot x, \qquad u = ax^{\frac{1}{2}},$$

are examples of such functions. In the first case,
$u = -\infty$, when $x = 0$;* in the second, $u = \infty$,
when $x = 0$; and in the third, B, and the succeeding
differential coefficients, become infinite, when $x = 0$.

TAYLOR'S THEOREM.

77. TAYLOR'S THEOREM explains the method of developing into a series any function of the sum or difference
of two independent variables.

78. Since the sum or difference of two independent
variables may always be denoted by a single letter, any
function of the form,

$$u' = f(x \pm y),$$

may be put under the form,

$$u' = f(z), \qquad \text{by making} \qquad z = x \pm y.$$

If we suppose z to be the abscissa, and u' the or·
dinate of a curve, and give to z an increment h, z will

* Bourdon, Art. **235**. University, Art. **186**. Legendre, Trig., Art. **22**.

become $z + h$. If we pass to consecutive values, $dz = dx$, and

$$\frac{du'}{dz} = \frac{du'}{dx} = \tan \alpha. \quad (\text{Art. } \mathbf{13.})$$

If we suppose x to remain constant, and y to receive the increment h, z will again become $z + h$, and when we pass to consecutive values,

$$\frac{du'}{dz} = \frac{du'}{dy} = \tan \alpha.$$

Hence, *in any function of the sum or difference of two independent variables, the partial differential coefficients are equal* (Art. **32**).

79. As an example, take,

$$u' = (x + y)^n.$$

If we suppose x to vary, the first partial differential coefficient is,

$$\frac{du'}{dx} = n(x + y)^{n-1}.$$

If we suppose y to vary, it is,

$$\frac{du'}{dy} = n(x + y)^{n-1};$$

and the same may be shown for the differential coefficients of the higher orders.

80. If any function of the form,

$$u' = f(x + y),$$

be developed into a series, it is plain that the series

must have terms containing the variables x and y, and that the constants, which enter into the given function, must also enter into the development. Let us then assume,

$$f(u') = f(x + y) = A + By^a + Cy^b + Dy^c + \&c. \quad (1.)$$

in which the terms are arranged according to the ascending powers of y, and in which A, B, C, D, &c., are independent of y, but functions of x, and arbitrary functions of all the constants which enter the primitive function. It is now required to find such values for the exponents a, b, c, &c., and for the coefficients A, B, C, D, &c., as shall render the development true for all possible values that may be attributed to x and y.

In the first place, there can be no negative exponents. For, if any term were of the form,

$$By^{-a},$$

it might be written,

$$\frac{B}{y^a},$$

and making $y = 0$, this term would become infinite, and we should have,

$$u' = f(x) = \infty,$$

which is absurd, since the function of x, which is independent of y, does not *necessarily* become infinite when $y = 0$.

The first term A, of the development, is the value which the primitive function u' assumes when we make $y = 0$.

If we designate this value by u, we shall have,

$$u = f(x).$$

If we differentiate Equation (1), under the supposition that x varies, the partial differential coefficient is,

$$\frac{du'}{dx} = \frac{dA}{dx} + \frac{dB}{dx}y^a + \frac{dC}{dx}y^b + \frac{dD}{dx}y^c + \&\text{c.} \ . \ (2.)$$

and if we differentiate, regarding y as a variable, the partial differential coefficient is,

$$\frac{du'}{dy} = aBy^{a-1} + bCy^{b-1} + cDy^{c-1} + \&\text{c.} \ . \ . \ (3.)$$

But these differential coefficients are equal to each other (Art. **78**); hence, the second members of Equations (2) and (3) are equal. Since the coefficients are independent of y, and the equality exists whatever be the value of y, it follows that the corresponding terms in each series will contain like powers of y, and that the coefficients of y in these terms will be equal.* Hence,

$$a - 1 = 0, \quad b - 1 = a, \quad c - 1 = b, \quad \&\text{c.,}$$

and consequently,

$$a = 1, \qquad b = 2, \qquad c = 3, \quad \&\text{c.}$$

Comparing the coefficients, we find,

$$B = \frac{dA}{dx}, \qquad C = \frac{1}{2}\frac{dB}{dx}, \qquad D = \frac{1}{3}\frac{dC}{dx}.$$

---

* Bourdon, Art. **195**. University, Art. **178**.

Since we have made,

$$f(x + y) = u', \qquad \text{and} \qquad f(x) = A = u,$$

we shall have,

$$A = u, \quad B = \frac{du}{dx}, \quad C = \frac{d^2u}{1.2\,dx^2}, \quad D = \frac{d^3u}{1.2.3\,dx^3},$$

and consequently,

$$u' = u + \frac{du}{dx}y + \frac{d^2u}{dx^2}\frac{y^2}{1.2} + \frac{d^3u}{dx^3}\frac{y^3}{1.2.3} + \&c.,$$

which is the formula of Taylor.

In this formula, u is what u' becomes, when $y = 0$; $\frac{du}{dx}$, what $\frac{du'}{dx}$ becomes when $y = 0$; $\frac{d^2u}{dx^2}$, what $\frac{d^2u'}{dx^2}$ becomes when $y = 0$; and similarly for the other coefficients.

1. Let it be required to develop

$$u' = f(x + y) = (x + y)^n,$$

by this formula.

We find,

$$u = x^n, \quad \frac{du}{dx} = n \cdot x^{n-1}, \quad \frac{d^2u}{dx^2} = n \cdot (n-1)x^{n-2} + \&c.;$$

hence,

$$u' = (x + y)^n = x^n + nx^{n-1}y + \frac{n(n-1)}{1.2}x^{n-2}y^2$$

$$+ \frac{n(n-1)(n-2)}{1.2.3}x^{n-3}y^3 + \&c.$$

SECTION V.

MAXIMA AND MINIMA.

81. A MAXIMUM value of a variable function is greater than the consecutive value which precedes, and the consecutive value which follows it.

A MINIMUM value of a variable function is less than the consecutive value which precedes, and the consecutive value which follows it.

If we denote any variable function by u, and the independent variable by x, every relation between u and x will be denoted by the co-ordinates of a curve whose equation is (Art. **10**),

$$u = f(x).$$

Let u' denote the consecutive ordinate which precedes u, and u'' the consecutive ordinate which follows it. Then, if u is a maximum,

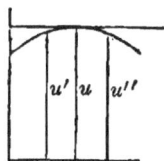

$$u > u', \qquad \text{and} \qquad u > u'';$$

the curve therefore ascends *just before* the ordinate reaches a maximum value, and descends *immediately afterwards;* hence, at the point of maximum, it is concave towards the axis of abscissas (Art. **73**).

Since the curve *ascends* just before the ordinate reaches the maximum value, the first differential coefficient will be positive; and since it then *descends*, the first differential coefficient will be negative immediately after the maximum value (Art. **72**). Hence, at the *point of maximum* value of the ordinate, the first differential coefficient will change its sign, and therefore passes through 0.

Since the curve is concave towards the axis of abscissas, the second differential coefficient is negative (Art. **73**); hence, the conditions of a maximum value of u are,

$$\frac{du}{dx} = 0, \quad \text{and} \quad \frac{d^2u}{dx^2}, \text{ negative.}$$

82. Denoting the consecutive ordinates, as before, by u', u, u'', if u is a minimum,

$$u < u', \quad \text{and} \quad u < u'';$$

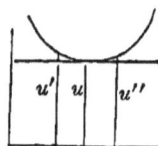

the curve, therefore, descends *just before* the ordinate reaches a minimum, and ascends *immediately afterwards;* hence, at the point of minimum, it is convex towards the axis of abscissas.

Since the curve *descends* just before the ordinate reaches the minimum value, the first differential coefficient will be negative; and since it then *ascends*, the first differential coefficient will be positive immediately after the minimum value (Art. **72**). Hence, at the point of *minimum value* of the ordinate, the first differential coefficient will change its sign, and therefore passes through 0.

Since the curve is convex towards the axis of abscissas, the second differential coefficient is positive (Art. **73**); hence, the conditions of a minimum value of u, are,

$$\frac{du}{dx} = 0, \quad \text{and} \quad \frac{d^2u}{dx^2}, \text{ positive.}$$

83. Hence, to find the maximum or minimum value of a function of a single variable:

1. *Find the first differential coefficient of the function, place it equal to 0, and determine the roots of the equation.*

2. *Find the second differential coefficient, and substitute each real root, in succession, for the variable in the second member of the equation; each root which gives a negative result, will correspond to a maximum value of the function, and each which gives a positive result will correspond to a minimum value.*

Point of inflection.

84. A POINT OF INFLECTION is a point at which a curve changes its curvature with respect to the axis of abscissas.

When a curve is concave towards the axis of abscissas, its second differential coefficient is negative (Art. **72**); when it is convex, the second differential coefficient is positive (Art. **72**): therefore, at the point where the curve changes its curvature, the second differential coefficient changes its sign, and consequently passes through zero.

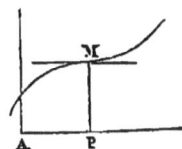

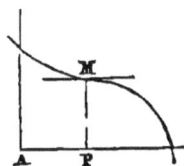

In the first figure, the second differential coefficient, at the point *M*, changes from negative to positive; in the second, from positive to negative. At the point *M*, in both figures, the first differential coefficient is equal to 0, and the tangent line separates the two branches

of the curve. When the second differential coefficient is
0, the ordinate at the point has neither a maximum nor
a minimum.

There are three consecutive points of the curve which
coincide with the tangent, at the point of inflection. This
is shown by the equality of the co-ordinates of the point
M (in the curve and tangent), and of the first and second differentials.

EXAMPLES.

1. To find the value of x which will render the function y a maximum or minimum in the equation of the
circle,

$$y^2 + x^2 = R^2. \qquad \frac{dy}{dx} = -\frac{x}{y};$$

making, $\qquad -\frac{x}{y} = 0, \qquad$ gives, $\qquad x = 0.$

The second differential coefficient is,

$$\frac{d^2y}{dx^2} = -\frac{x^2 + y^2}{y^3}. \qquad \text{When,} \quad x = 0, \quad y = R;$$

hence, $\qquad \dfrac{d^2y}{dx^2} = -\dfrac{1}{R},$

which being negative, y is a maximum for R positive.

2. Find the values of x which render the function y
a maximum or minimum in the equation,

$$y = a - bx + x^2. \qquad \text{Differentiating,}$$

$$\frac{dy}{dx} = -b + 2x, \qquad \text{and} \qquad \frac{d^2y}{dx^2} = 2;$$

and making, $-b + 2x = 0,$

gives, $x = \dfrac{b}{2}.$

Since the second differential coefficient is. positive, this value of x will render y a minimum. The minimum value of y is found by substituting the value of x, in the primitive equation. It is,

$$y = a - \frac{b^2}{4}.$$

3. Find the value of x which will render the function u a maximum or minimum in the equation,

$$u = a^4 + b^3x - c^2x^2.$$

$$\frac{du}{dx} = b^3 - 2c^2x, \qquad \text{hence,} \qquad x = \frac{b^3}{2c^2},$$

and $$\frac{d^2u}{dx^2} = - 2c^2;$$

hence, the function is a maximum, and the maximum value is,

$$u = a^4 + \frac{b^6}{4c^2}.$$

4. Let us take the function,

$$u = 3a^2x^3 - b^4x + c^5.$$

We find, $\dfrac{du}{dx} = 9a^2x^2 - b^4,$ and $x = \pm \dfrac{b^2}{3a}.$

The second differential coefficient is,

$$\frac{d^2u}{dx^2} = 18a^2x.$$

Substituting the plus root of x, we have,

$$\frac{d^2u}{dx^2} = + 6ab^2,$$

which gives a minimum, and substituting the negative root, we have,

$$\frac{d^2u}{dx^2} = - 6ab^2,$$

which gives a maximum.

The minimum value of the function is,

$$u = c^5 - \frac{2b^6}{9a};$$

and the maximum value,

$$u = c^5 + \frac{2b^6}{9a}.$$

5. Find the values of x, which make u a maximum or minimum in the equation,

$$u = x^5 - 5x^4 + 5x^3 - 1.$$

Ans. $\begin{cases} x = 1, \text{ a maximum.} \\ x = 3, \text{ a minimum.} \end{cases}$

6. Find the values of x, which make u a maximum or minimum in the equation,

$$u = x^3 - 9x^2 + 15x - 3.$$

Ans. $\begin{cases} x = +1, \text{ a maximum.} \\ x = +5, \text{ a minimum.} \end{cases}$

7. Find the values of x, which make u a maximum or minimum in the equation,

$$u = x^3 - 3x^2 + 3x + 7.$$

Ans. There is no such value of x, since the second differential coefficient reduces to 0, for $x = 1$; hence, only one condition of a maximum or minimum is fulfilled.*

85. NOTES. 1. In applying the preceding rules to practical examples, we first find an expression for the function which is to be made a maximum or minimum.

2. If in such expression, a constant quantity is found as a *factor*, it may be omitted in the operation; for the product will be a maximum or a minimum when the variable factor is a maximum or minimum.

3. Any value of the independent variable which renders a function a maximum or a minimum, will render any power or root of that function, a maximum or minimum; hence, we may square both members of an equation to free it of radicals, before differentiating.

8. To find the maximum rectangle which can be inscribed in a given triangle.

Let b denote the base of the triangle, h the altitude, y the base of the rectangle, and x its altitude. Then,

$$u = xy = \text{the area of the rectangle.}$$

But, $$b : h :: y : h - x;$$

hence, $$y = \frac{bh - bx}{h},$$

---

* We have limited the discussion to a single class of maxima and minima, viz.: that in which the first differential coefficient of the function is 0, and the second negative or positive.

and consequently,

$$u = \frac{bhx - bx^2}{h} = \frac{b}{h}(hx - x^2);$$

and omitting the constant factor $\frac{b}{h}$, we may write,

$$u' = hx - x^2;$$

for, the value of x, which makes u' a maximum, will make u a maximum (Art. 85); hence,

$$\frac{du'}{dx} = h - 2x, \qquad \text{or,} \qquad x = \frac{h}{2};$$

therefore, the altitude of the rectangle is equal to half the altitude of the triangle; and since,

$$\frac{d^2u'}{dx^2} = -2,$$

the area is a maximum (Art. 81).

9. What is the altitude of a cylinder inscribed in a given cone, when the volume of the cylinder is a maximum?

Suppose the cylinder to be inscribed, as in the figure, and let

$AB = a,\ BC = b,\ AD = x,\ ED = y;$

then, $BD = a - x =$ altitude of the cylinder, and

$\pi y^2(a - x)^* =$ volume $= v$ .. (1.)

From the similar triangles AED and ACB, we have,

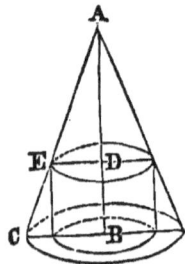

---

* Legendre, Bk. VIII. Prop. 2.

$$x : y :: a : b; \quad \text{whence,} \quad y = \frac{bx}{a}.$$

Substituting this value in Equation (1), we have,

$$v = \frac{\pi b^i}{a^2} x^2 (a - x).$$

Omitting the constant factor $\frac{\pi b^2}{a^2}$, we may write,

$$v' = x^2 (a - x);$$

for, the conditions which will make v' a maximum, will also make v a maximum (Art. 85).

By differentiating, we have,

$$\frac{dv'}{dx} = 2ax - 3x^2.$$

Placing, $\qquad\qquad 2ax - 3x^2 = 0,$

we have, $\qquad x = 0, \quad$ and $\quad x = \frac{2}{3}a.$

But, $\qquad\qquad \frac{d^2v'}{dx^2} = 2a - 6x = -2a.$

Hence, the cylinder is a maximum, when its altitude is one-third the altitude of the cone.

10. What is the altitude of a cone inscribed in a given sphere, when the volume is a maximum?

Denote the radius of the given sphere by r, and the centre by C. Let A be the vertex of the required cone, BD the radius of its base, which denote by y, and denote the altitude AB by x. Then,

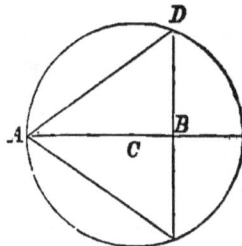

$$y^2 = 2rx - x^2;^*$$

and if we denote the volume of the cone by v,

$$v = \tfrac{1}{3}\pi x(2rx - x^2) = \tfrac{1}{3}\pi(2rx^2 - x^3).\dagger$$

Omitting the constant factor $\tfrac{1}{3}\pi$, we have,

$$\frac{dv'}{dx} = 4rx - 3x^2; \qquad \text{hence,}$$

$$4rx - 3x^2 = 0, \qquad \text{and} \qquad x = \frac{4}{3}r;$$

that is, the altitude of the cone is four-thirds of the radius.

11. What is the altitude of a cone inscribed in a sphere when the convex surface is a maximum? *Ans.* $\dfrac{4}{3}r$.

12. What is the length of the axis of a maximum parabola which can be cut from a given right cone with a circular base?

Let BAC be a section of the cone by a plane passed through the axis; and FDG a parabola made by a plane parallel to the element BA.

Denote BC by b, AB by a, and CE by x; then, $BE = b - x$, and FE, the common ordinate of the circle and parabola, is equal to $\sqrt{bx - x^2}.\ddagger$

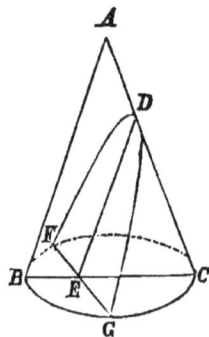

---

* An. G., Bk. II. Art. 4—8.   † Leg., Bk. VIII. Prop. 5.
‡ Legendre, Bk. IV. Prop. 23. Cor. 2.

By similar triangles, we have,

$$b : a :: x : \frac{ax}{b} = DE.$$

Hence, the area of the parabola (Art. **59**) is,

$$u = \frac{2}{3} \frac{ax}{b} \sqrt{bx - x^2}.$$

Omitting the constant factors, and remembering that the same value of x, which renders u a maximum, will render its square a maximum (Art. **79**), and designating by u' the new function, we have,

$$u' = x^2(bx - x^2) = bx^3 - x^4, \qquad \text{and}$$

$$\frac{du'}{dx} = 3bx^2 - 4x^3; \quad \text{or,} \quad x = \frac{3}{4}b; \quad \text{and} \quad DE = \frac{3}{4}AB.$$

that is, the axis of the maximum parabola is three-fourths the slant height of the cone.

13. What is the altitude of the maximum rectangle which can be inscribed in a given parabola?

Ans. Two-thirds of the axis.

14. What are the sides of the maximum rectangle inscribed in a given circle?

Ans. A square whose side is $r\sqrt{2}$.

15. A cylindrical vessel, open at top, is to contain a given quantity of water. What is the relation between the radius of the base and the altitude, when the interior surface is a minimum?

Ans. Altitude = radius of base.

16. Required the maximum right-angled triangle which ʼan be constructed on a given line, as a hypothenuse?

Ans. When it is isosceles.

17. Required the least triangle which can be formed by the two radii, produced, and a tangent line to the quadrant of a given circle? *Ans.* When it is isosceles.

18. What is the altitude of the maximum cylinder which can be inscribed in a given paraboloid?

Ans. Half the axis.

19. What is the altitude of a cylinder inscribed in a given sphere when its convex surface is a maximum?

Ans. $r\sqrt{2}$.

20. What is the altitude of a cylinder inscribed in a given sphere, when its volume is a maximum?

Ans. $\dfrac{2r}{\sqrt{3}}$.

21. Required the base of the maximum rectangle which can be inscribed in a given ellipse whose semi-axes are A and B. *Ans.* $A\sqrt{2}$.

22. A rectangular sheep-fold, *to contain a given area,* is to be built against a wall. Required the ratio of the least side to the larger, so that the cost shall be a minimum. *Ans.* 2.

23. To circumscribe a given circle whose radius is r, by an isosceles triangle whose area shall be a minimum.

Ans. Perpendicular to base $= 3r$

SECTION VI.

Differentials of Exponential and Logarithmic functions.

86. An Exponential function is one in which the in-dependent variable enters as an exponent; as,

$$u = a^x \quad . \quad . \quad . \quad . \quad . \quad . \quad (1.)$$

If, in a function of this form, we give to x an incre ment h, we have,

$$u' = a^{x+h} = a^x a^h \quad . \quad . \quad . \quad . \quad (2.)$$

Subtracting Equation (1) from (2), member from member, we have,

$$u' - u = a^x a^h - a^x = a^x(a^h - 1) ;$$

whence,
$$\frac{u' - u}{a^x} = a^h - 1 \quad . \quad . \quad . \quad . \quad . \quad (3.)$$

Put, $a = 1 + b$, and develop by the binomial formula; we then have,

$$a^h = (1 + b)^h = 1 + hb + \frac{h}{1}\left(\frac{h-1}{2}\right)b^2 + \frac{h}{1}\left(\frac{h-1}{2}\right)\left(\frac{h-2}{3}\right)b^3$$

$$+ \frac{h}{1}\left(\frac{h-1}{2}\right)\left(\frac{h-2}{3}\right)\left(\frac{h-3}{4}\right)b^4 + \&c.$$

Substituting this value of a^h, in Equation (3), and dividing by h, we have,

$$\frac{u' - u}{a^x h} = b + \left(\frac{h-1}{2}\right)b^2 + \left(\frac{h-1}{2}\right)\left(\frac{h-2}{3}\right)b^3$$

$$+ \left(\frac{h-1}{2}\right)\left(\frac{h-2}{3}\right)\left(\frac{h-3}{4}\right)b^4 + \&c.$$

If we now pass to consecutive values, by making h numerically equal to 0, we have,

$$\frac{du}{a^x dx} = b - \frac{b^2}{2} + \frac{b^3}{3} - \frac{b^4}{4} + \frac{b^5}{5} - \&c. ;$$

and putting for b its value, $a - 1$, we have,

$$\frac{du}{a^x dx} = a - 1 - \frac{(a-1)^2}{2} + \frac{(a-1)^3}{3} - \frac{(a-1)^4}{4} + \&c. \quad (4.)$$

Denoting the second member of Equation (4) by k, we have,

$$\frac{du}{a^x dx} = k, \quad \text{or,} \quad du = da^x = a^x k dx \quad . \quad . \quad (5.)$$

that is, *the differential of a function of the form a^x, is equal to the function, into a constant quantity k, dependent on a, into the differential of the exponent.*

Relation between a and k.

87. The relation between a and k is very peculiar, and may be determined by Maclaurin's Formula,

$$u = a^x = \left(u\right) + \left(\frac{du}{dx}\right)x + \frac{1}{1.2}\left(\frac{d^2u}{dx^2}\right)x^2 + \frac{1}{1.2.3}\left(\frac{d^3u}{dx^3}\right)x^3$$

$$+ \&c. \quad . \quad . \quad . \quad . \quad . \quad (6.)$$

First, if we make $x = 0$, the function $a^x = 1 = \left(u\right)$. The successive differential coefficients are found from Equation (5); viz. :

$$\frac{du}{dx} = a^x k, \quad \text{and} \quad \left(\frac{du}{dx}\right) = k \; ;$$

$$d\left(\frac{du}{dx}\right) = \frac{d^2u}{dx} = da^x k = a^x k^2 dx \; ; \quad \text{hence,}$$

$$\frac{d^2u}{dx^2} = a^x k^2, \quad \text{and} \quad \left(\frac{d^2u}{dx^2}\right) = k^2 \; ;$$

$$\frac{d^3u}{dx^3} = a^x k^3, \quad \text{and} \quad \left(\frac{d^3u}{dx^3}\right) = k^3,$$

$$\&\text{c.,} \qquad \&\text{c.,} \qquad \&\text{c.}$$

Substituting these values in Equation (6), we have,

$$u = a^x = 1 + \frac{kx}{1} + \frac{k^2x^2}{1.2} + \frac{k^3x^3}{1.2.3} + \&\text{c.}$$

If we make $x = \dfrac{1}{k}$, we shall have,

$$a^{\frac{1}{k}} = 1 + \frac{1}{1} + \frac{1}{1.2} + \frac{1}{1.2.3} + \&\text{c.} \; ;$$

designating the second member of the equation by e, and employing twelve terms of the series, we find,

$$e = 2.7182818\ldots. \; ;$$

hence, $a^{\frac{1}{k}} = e$,   therefore, $a = e^k$ . . **(7.)**

Equation (7) expresses the relation between a and k.

A system of logarithms, called the Naperian system, has been constructed, whose base is, $e = 2.7182818\ldots$ This, and the common system, whose base is 10, are the only systems in use. The logarithms, in the Naperian system, are denoted by l, and in the common system by log. We see from Equation (7), that k is the Naperian logarithm of the number a. If we take the common logarithms of both members of Equation (7), we shall have,

$$\log a = k \log e \quad \ldots \quad \ldots \quad (8.)$$

The common logarithm of $e = \log 2.7182818\ldots$ $= .434284482\ldots$, is called the modulus of the common system, and is denoted by M. Hence, *if we have the Naperian logarithm of a number, we can find the common logarithm of the same number by multiplying by the modulus.*

If, in Equation (8), we make $a = 10$, we have,

$$1 = k \log e; \quad \text{or,} \quad \frac{1}{k} = \log e = M;$$

that is, *the modulus of the common system is also equal to 1, divided by the Naperian logarithm of the common base.*

88. From Equation (5), we have,

$$\frac{du}{u} = \frac{da^z}{a^z} = kdx.$$

If we make $a = 10$, the base of the common system, $x = \log u$, and

$$dx = \frac{du}{u} \times \frac{1}{k} = \frac{du}{u} \times M;$$

that is, *the differential of a common logarithm of a quantity is equal to the differential of the quantity divided by the quantity into the modulus.*

89. If we make $a = e$, the base of the Naperian system, x becomes the Naperian logarithm. of u, and k becomes 1: see Equation (7); hence, $M = 1$; and

$$dx = \frac{du}{a^x};$$

that is, *the differential of a Naperian logarithm of a quantity is equal to the differential of the quantity divided by the quantity; and in this system, the modulus is* 1.

90. Having found that k is the Naperian logarithm of a, we have from Equation (5),

$$du = a^x \, l \, a \, dx;$$

that is, *the differential of a function of the form a^x, is equal to the function, into the Naperian logarithm of the base a, into the differential of the exponent.*

EXAMPLES.

1. Find the differential of $u = a^x$.
$$du = a^x \, l \, a \, dx.$$

2. Find the differential of $u = l \, x$.
$$du = \frac{dx}{x} = x^{-1} dx.$$

NOTE. This case would seem to admit of integration by the rule of Art. **35**; but that rule applies to alge-

braic functions only, and this form is derived from a
transcendental function.

3. Find the differential of $u = y^x$.

$$l\,u = x\,l\,y; \text{ hence,}$$

$$\frac{du}{u} = x\frac{dy}{y} + l\,y\,dx; \text{ hence,}$$

by clearing of fractions, and reducing,

$$du = xy^{x-1}dy + y^x l\,y\,dx;$$

that is, equal to the sum of the partial differentials
(Art. 32).

4. Find by logarithms the differential of $u = xy$.

$$l\,u = l\,x + l\,y;^* \text{ hence,}$$

$$\frac{du}{u} = \frac{dx}{x} + \frac{dy}{y}; \text{ and by reducing,}$$

$$du = ydx + xdy \quad (\text{Art. } 27).$$

5. Find by logarithms the differential of $u = \frac{x}{y}$.

$$l\,u = l\,x - l\,y;\dagger \text{ hence, by differentiating,}$$

$$\frac{du}{u} = \frac{dx}{x} - \frac{dy}{y}; \text{ and by reducing,}$$

$$du = \frac{ydx - xdy}{y^2} \quad (\text{Art. } 29).$$

---

* Bourdon, Art. 230. University, Art. 185.
† Bourdon, Art. 231. University, Art. 185.

6. Find the differential of $u = l\left(\dfrac{x+a}{a-x}\right)$.

$$du = \frac{2a\,dx}{a^2 - x^2}.$$

7. Find the differential of $u = l\left(\dfrac{\dot{x}}{\sqrt{a^2 + x^2}}\right)$.

$$du = \frac{a^2 dx}{x(a^2 + x^2)}.$$

8. Find the differential of $u = (a^z + 1)^2$.

$$du = 2a^z(a^z + 1)\,l\,a\,dx.$$

9. Find the differential of $u = \dfrac{a^z - 1}{a^z + 1}$.

$$du = \frac{2a^z\,l\,a\,dx}{(a^z + 1)^2}.$$

10. Find the differential of $u = \dfrac{a^z}{x^z} = \left(\dfrac{a}{x}\right)^z$.

$$du = \left(\frac{a}{x}\right)^z\left(l\,\frac{a}{x} - 1\right)dx.$$

Differential forms which have known integrals.

91. If we have a differential in a fractional form, in which the numerator is the differential of the denominator, we know that the integral is the Naperian logarithm of the denominator (Art. **89**). It frequently happens, however, that we have to deal with fractional differentials which are not of this form, but which, by certain algebraic artifices, may be reduced to it. We shall give a few examples of such reductions.

Form 1.
$$\int \frac{dx}{\sqrt{x^2 \pm a^2}}.$$

Put $x^2 \pm a^2 = v^2$;   then, $xdx = vdv.$

Add vdx to both members; then,

$$xdx + vdx = vdx + vdv;\quad \text{hence,}$$

$$(x + v)dx = v(dx + dv);\quad \text{whence,}$$

$$\frac{dx + dv}{x + v} = \frac{dx}{v} = \frac{dx}{\sqrt{x^2 \pm a^2}};\quad \text{hence,}$$

$$\int \frac{dx + dv}{x + v} = \int \frac{dx}{\sqrt{x^2 \pm a^2}}.$$

But in the first member, the numerator is the differential of the denominator; hence,

$$\int \frac{dx}{\sqrt{x^2 \pm a^2}} = l(x + v) = l(x + \sqrt{x^2 \pm a^2}).$$

Form 2.
$$\int \frac{dx}{\sqrt{x^2 \pm 2ax}}.$$

Put $\sqrt{x^2 \pm 2ax} = {'}v$;   then, $x^2 \pm 2ax = v^2.$

Adding a^2 to both members, and extracting the square
root,

$$x \pm a = \sqrt{v^2 + a^2};\quad \text{hence,}\quad dx = \frac{vdv}{\sqrt{v^2 + a^2}},$$

and
$$\frac{dx}{\sqrt{x^2 \pm 2ax}} = \frac{dv}{\sqrt{v^2 + a^2}}.$$

But from the first form,

$$\int \frac{dv}{\sqrt{v^2 + a^2}} = l\left(v + \sqrt{v^2 + a^2}\right).$$

Substituting for v its value, and for $\sqrt{v^2 + a^2}$, its value,

$$\int \frac{dx}{\sqrt{x^2 \pm 2ax}} = l\left(x \pm a + \sqrt{x^2 \pm 2ax}\right).$$

Form 3. $\dfrac{2adx}{a^2 - x^2};$ or, $\dfrac{2adx}{x^2 - a^2}.$

Since, $\dfrac{2adx}{a^2 - x^2} = \dfrac{2adx}{(a+x)(a-x)} = \dfrac{dx}{a+x} + \dfrac{dx^*}{a-x}.$

$$\int\left(\frac{dx}{a+x} + \frac{dx}{a-x}\right) = \int \frac{dx}{a+x} + \int \frac{dx}{a-x}$$

$$\int \frac{2adx}{a^2 - x^2} = l(a+x) - l(a-x) = l\left(\frac{a+x}{a-x}\right).$$

Also, $$\int \frac{2adx}{x^2 - a^2} = l\left(\frac{x-a}{x+a}\right).$$

(See Example 6, page 126.)

Form 4. $\dfrac{2adx}{x\sqrt{a^2 \pm x^2}}.$

Put $\sqrt{a^2 + x^2} = v;$ whence, $a^2 + x^2 = v^2;$ hence,

$x^2 = v^2 - a^2,$ and $xdx = vdv,$ or, $dx = \dfrac{vdv}{x}.$

---

* University, Art. **180**. (See Art. **158**.)

Multiply both members by $\dfrac{2a}{x\sqrt{a^2 + x^2}}$; we have,

$$\int \frac{2a\,dx}{x\sqrt{a^2 + x^2}} = \int \frac{2a\,dv}{v^2 - a^2} = l\left(\frac{v - a}{v + a}\right); \quad \text{hence,}$$

$$\int \frac{2a\,dx}{x\sqrt{a^2 + x^2}} = l\left(\frac{\sqrt{a^2 + x^2} - a}{\sqrt{a^2 + x^2} + a}\right).$$

In like manner we should find,

$$\int \frac{2a\,dx}{x\sqrt{a^2 - x^2}} = l\left(\frac{a - \sqrt{a^2 - x^2}}{a + \sqrt{a^2 - x^2}}\right).$$

Form 5. $\qquad\qquad \displaystyle\int \frac{x^{-2}\,dx}{\sqrt{a^2 + x^{-2}}}.$

Put $\quad \dfrac{1}{x} = v;\qquad$ then, $\qquad x^{-2}\,dx = -\,dv;\qquad$ and

$$\frac{x^{-2}\,dx}{\sqrt{a^2 + x^{-2}}} = \frac{-\,dv}{\sqrt{a^2 + v^2}}; \quad \text{first form.}$$

$$\int \frac{x^{-2}\,dx}{\sqrt{a^2 + x^{-2}}} = -\,l\left(v + \sqrt{a^2 + v^2}\right)$$

$$= -\,l\left(\frac{1}{x} + \sqrt{a^2 + \frac{1}{x^2}}\right)$$

$$= -\,l\left(\frac{1 + \sqrt{1 + a^2 x^2}}{x}\right).$$

TABLE OF FORMS.

1. $\int a^x \, l \, a \, dx = a^x$ . . . . (Ex. 1.)

2. $\int \dfrac{dx}{x} = \int dx \, x^{-1} = l \, x$ . . . . . (Ex. 2.)

3. $\int (x y^{x-1} dy + y^x \, l \, y \times dx) = y^x$. (Ex. 3.)

4. $\int \dfrac{dx}{\sqrt{x^2 \pm a^2}} = l\left(x + \sqrt{x^2 \pm a^2}\right).$ (Form 1.)

5. $\int \dfrac{dx}{\sqrt{x^2 \pm 2ax}} = l\left(x \pm a + \sqrt{x^2 \pm 2ax}\right).$ (2.)

6    $\int \dfrac{2a \, dx}{a^2 - x^2} = l\left(\dfrac{a + x}{a - x}\right).$ (Form 3.)

7. $\int \dfrac{2a \, dx}{x^2 - a^2} = l\left(\dfrac{x - a}{x + a}\right).$ (Form 3.)

8. $\int \dfrac{2a \, dx}{x\sqrt{a^2 + x^2}} = l\left(\dfrac{\sqrt{a^2 + x^2} - a}{\sqrt{a^2 + x^2} + a}\right).$ (Form 4.)

9. $\int \dfrac{2a \, dx}{x\sqrt{a^2 - x^2}} = l\left(\dfrac{a - \sqrt{a^2 - x^2}}{a + \sqrt{a^2 - x^2}}\right).$ (Form 4.)

10. $\int \dfrac{x^{-2} dx}{\sqrt{a^2 + x^{-2}}} = - l\left(\dfrac{1 + \sqrt{1 + a^2 x^2}}{x}\right).$ (5.)

CIRCULAR FUNCTIONS.

92. LET O be the centre of a circle, A the origin of arc, and BC, DH any two consecutive ordinates. Draw BE parallel to OA; draw the radius OB, and denote it by 1. Denote the arc AB by z, and suppose z to be the independent variable. Then, BD will be the differential of the arc AB; ED, the differential of the sine, and EB the differential of the cosine, which will be negative, since it is a decreasing function of the arc ('Art. **19**).

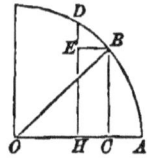

93. Since the triangles OBC and DEB, have their sides respectively perpendicular to each other, they will be similar;* hence,

$$OB : OC :: BD : DE; \text{ or,}$$

$$1 : \cos z :: dz : d \sin z; \text{ whence,}$$

$$d \sin z = \cos z \, dz \quad \ldots \ldots \quad (1.)$$

94. Again, $1 : \sin z :: dz : - d \cos z;$ whence,

$$d \cos z = - \sin z \, dz \quad \ldots \ldots \quad (2.)$$

95. Since,

$$\cos z = 1 - \text{ver-sin } z, \qquad d \cos z = - d \text{ ver-sin } z;$$

hence, $d \text{ ver-sin } z = \sin z \, dz \quad \ldots \ldots \quad (3.)$

---

* Legendre, Bk. IV. Prop. 21.

96. Again, $\tan z = \dfrac{\sin z}{\cos z}$;   hence,

$$d \tan z = \frac{\cos z \, d \sin z - \sin z \, d \cos z}{\cos^2 z} \text{ (Art. 29).}$$

Substituting for $d \sin z$ and $d \cos z$, their values from Equations (1) and (2), we have,

$$d \tan z = \frac{dz}{\cos^2 z} \quad \cdots \cdots \quad (4.)$$

By similar processes, we can find the differentials of the co-versed-sine, cotangent, secant, and cosecant, in terms of the other functions and the differential of z.

97. Denote the sine of the arc AB by y, its cosine by x, its versed sine by v, and its tangent by t. If we regard each of these as the independent variable, and the arc z as the common function, and find the values of z from Equations (1), (2), (3), and (4), we shall have,

When radius $= 1$,

$$dz = \frac{dy}{\sqrt{1 - y^2}} \cdots \cdots (5.)$$

$$dz = -\frac{dx}{\sqrt{1 - x^2}} \cdots (6.)$$

$$dz = \frac{dv}{\sqrt{2v - v^2}} \cdots (7.)$$

$$dz = \frac{dt}{1 + t^2} \cdots \cdots (8.)$$

When radius $= r$,

$$dz = \frac{r \, dy}{\sqrt{r^2 - y^2}} \cdots (9.)$$

$$dz = -\frac{r \, dx}{\sqrt{r^2 - x^2}} \cdot (10.)$$

$$dz = \frac{r \, dv}{\sqrt{2rv - v^2}} \cdots (11.)$$

$$dz = \frac{r^2 \, dt}{r^2 + t^2} \cdots \cdots (12.)$$

The differential of the arc, in terms of either of the other functions is easily found.

98. The following notation is employed to designate an arc by means of any one of its functions.

$\sin^{-1}u$, denotes the arc of which u is the sine,

$\cos^{-1}u$, denotes the arc of which u is the cosine,

$\tan^{-1}u$, denotes the arc of which u is the tangent,

&c.,   &c.,   &c.

If we denote the sine of an arc by $\dfrac{u}{a}$, instead of y, as in Equation (5), we shall have,

$$y = \frac{u}{a}, \qquad dy = \frac{du}{a}, \qquad \text{and} \qquad z = \sin^{-1}\frac{u}{a}.$$

Substituting these values in Equation (5), we have,

$$dz = \frac{du}{\sqrt{a^2 - u^2}} \quad \cdots \cdots \quad (13.)$$

Denoting the cosine of the arc by $\dfrac{u}{a}$, and making like substitutions in Equation (6), we have,

$$dz = \frac{-du}{\sqrt{a^2 - u^2}} \quad \cdots \cdots \quad (14.)$$

Denoting the ver-sine of the arc by $\dfrac{u}{a}$, and making like substitutions in Equation (7), we have,

$$dz = \frac{du}{\sqrt{2au - u^2}} \quad \cdots \cdots \quad (15.)$$

Denoting the tangent of an arc by $\dfrac{u}{a}$, we have from Equation (8),

$$dz = \frac{adu}{a^2 + u^2} \cdot \quad \cdots \quad (16.)$$

EXAMPLES.

1. Differentiate the function,

$$z = \cos^{-1}\left(u\sqrt{1 - u^2}\right).$$

$$dz = \frac{(-1 + 2u^2)du}{\sqrt{(1 - u^2 + u^4)(1 - u^2)}}\cdot$$

2. Differentiate the function,

$$z = \sin^{-1}\left(2u\sqrt{1 - u^2}\right). \qquad dz = \frac{2du}{\sqrt{1 - u^2}}\cdot$$

3. Differentiate the function,

$$z = \tan^{-1}\frac{x}{y}, \qquad dz = \frac{ydx - xdy}{y^2 + x^2}\cdot$$

4. Differentiate the function,

$$z = \cos x^{\sin x}.$$

Make, $\cos x = u$,   and $\sin x = y$;

then, $z = u^y$,   and,   (Art. **90**),

$$dz = u^y\, l\, u\, dy + yu^{y-1}\, du;$$

also, $du = -\sin x\, dx$,   and $dy = \cos x\, dx$;

hence, $dz = u^v \left(l u \, dy + \dfrac{y}{u} \, du \right)$

$$= \cos x^{\sin x} \left(l \cos x \cos x - \frac{\sin^2 x}{\cos x} \right) dx.$$

Differential forms which have known Integrals.

99. The first four equations in Art. **92** furnish us four forms, by taking the integrals of both members. Equations (5), (6), (7), and (8), are of the same form as Equations (9), (10), (11), and (12), except that the radius is 1 in the first set, and r in the second; hence, the arc z, in each equation of the second set, is r times as great as in the corresponding equation of the first set.*

Forms (13), (14), (15), and (16), are modified forms of (5), (6), (7), and (8). They differ from them only in the symbol by which the function of the arc is denoted.

TABLE OF FORMS.

1. $\displaystyle\int \cos z \, dz \quad = \sin z + C.$

2. $\displaystyle\int - \sin z \, dz \ = \cos z + C.$

3. $\displaystyle\int \sin z \, dz \quad = \text{ver-sin } z + C.$

4. $\displaystyle\int \frac{dz}{\cos^2 z} \quad = \tan z + C.$

5. $\displaystyle\int \frac{dy}{\sqrt{1 - y^2}} \ = \sin^{-1} y + C.$

---

* Leg., Trig. Art. 30.

6. $\displaystyle\int \frac{-\,dx}{\sqrt{1-x^2}} \;=\; \cos^{-1}x + C.$

7. $\displaystyle\int \frac{dv}{\sqrt{2v-v^2}} \;=\; \text{ver-sin}^{-1}v + C.$

8. $\displaystyle\int \frac{dt}{1+t^2} \;=\; \tan^{-1}t + C.$

9. $\displaystyle\int \frac{r\,dy}{\sqrt{r^2-y^2}} \;=\; \sin^{-1}y + C.$

10. $\displaystyle\int \frac{-\,r\,dx}{\sqrt{r^2-x^2}} \;=\; \cos^{-1}x + C.$

11. $\displaystyle\int \frac{r\,dv}{\sqrt{2rv+v^2}} \;=\; \text{ver-sin}^{-1}v + C.$

12. $\displaystyle\int \frac{r^2\,dt}{r^2+t^2} \;=\; \tan^{-1}t + C.$

To radius r.

13. $\displaystyle\int \frac{du}{\sqrt{a^2-u^2}} \;=\; \sin^{-1}\frac{u}{a} + C.$

14. $\displaystyle\int \frac{-\,du}{\sqrt{a^2-u^2}} \;=\; \cos^{-1}\frac{u}{a} + C.$

15. $\displaystyle\int \frac{du}{\sqrt{2au-u^2}} \;=\; \text{ver-sin}^{-1}\frac{u}{a} + C.$

16. $\displaystyle\int \frac{a\,du}{a^2+u^2} \;=\; \tan^{-1}\frac{u}{a} + C.$

Applications.

100. We may readily find the relation between the diameter and the circumference of a circle from either of the first four equations of Art. **97.**

1. To find this ratio from Equation (5), which is,

$$dz = \frac{dy}{\sqrt{1-y^2}}; \quad \text{or,} \quad \frac{dz}{dy} = \frac{1}{\sqrt{1-y^2}} = (1-y^2)^{-\frac{1}{2}}.$$

Developing by the Binomial Formula, we have,

$$\frac{dz}{dy} = 1 + \frac{1}{2}y^2 + \frac{1.3}{2.4}y^4 + \frac{1.3.5}{2.4.6}y^6 + \&c.; \quad \text{whence,}$$

$$dz = dy + \frac{1}{2}y^2 dy + \frac{1.3}{2.4}y^4 dy + \frac{1.3.5}{2.4.6}y^6 dy + \&c.$$

$$\int dz = z = y + \frac{1}{2.3}y^3 + \frac{1.3}{2.4.5}y^5 + \frac{1.3.5}{2.4.6.7}y^7 + \&c.$$

If we make $z = 30°$, of which the sine, y is $\frac{1}{2}$,[*] we have,

$$30° = \frac{1}{2} + \frac{1}{2.3.2^3} + \frac{1.3}{2.4.5.2^5} + \frac{1.3.5}{2.4.6.7.2^7} + \&c.$$

By multiplying both members of the equation by 6, and taking twelve terms of the series, we have,

$$180° = \pi = 3.1415924,$$

which is true to the last place, which should be 6.

---

[*] Legendre, Trig. Art. **64.**

2. Find the ratio from Equation (8), which is,

$$dz = \frac{dt}{1 + t^2}; \quad \text{or,} \quad \frac{dz}{dt} = \frac{1}{1 + t^2} = (1 + t^2)^{-1}.$$

Developing by the Binomial Formula, we have,

$$\frac{dz}{dt} = 1 - t^2 + t^4 - t^6 + t^8 - \&c. ; \quad \text{whence,}$$

$$dz = dt - t^2 dt + t^4 dt - t^6 dt + t^8 dt - \&c.$$

$$\int dz = z = \tan^{-1} t = t - \frac{t^3}{3} + \frac{t^5}{5} - \frac{t^7}{7} + \frac{t^9}{9} - \&c.$$

This series is not sufficiently converging. To find the value of the arc in a more converging series, we employ the following property of two arcs, viz.:

Four times the arc whose tangent is $\frac{1}{5}$, exceeds the arc of 45° by the arc whose tangent is $\frac{1}{239}$. *

---

* Let a denote the arc whose tangent is $\frac{1}{5}$. Then, Leg., Trig. Art. **36.**,

$$\tan 2a = \frac{2 \tan a}{1 - \tan^2 a} = \frac{5}{12},$$

$$\tan 4a = \frac{2 \tan 2a}{1 - \tan^2 2a} = \frac{120}{119}.$$

The last number being greater than 1, shows that the arc 4a exceeds 45°. Making,

$$4a = A, \qquad 45° = B,$$

But, $\quad \tan^{-1}\left(\dfrac{1}{5}\right) = \dfrac{1}{5} - \dfrac{1}{3.5^3} + \dfrac{1}{5.5^5} - \dfrac{1}{7.5^7} + \&c.,$

$\tan^{-1}\left(\dfrac{1}{239}\right) = \dfrac{1}{239} - \dfrac{1}{3(239)^3} + \dfrac{1}{5(239)^5} - \dfrac{1}{7(239)^7} + \&c.$

hence

$$\text{arc } 45° = \left\{ \begin{array}{l} 4\left(\dfrac{1}{5} - \dfrac{1}{3.5^3} + \dfrac{1}{5.5^5} - \dfrac{1}{7.7^7} +\right) \\[3mm] -\left(\dfrac{1}{239} - \dfrac{1}{3(239)^3} + \dfrac{1}{5(239)^5} - \dfrac{1}{7(239)^7} +\right) \end{array} \right\}$$

Multiplying both members by 4, we find,

$$180° = \pi = 3.141592653.$$

---

the difference, $4a - 45° = A - B = b$, will have for its tangent,

$$\tan b = \tan (A - B) = \frac{\tan A - \tan B}{1 + \tan A \tan B} = \frac{1}{239};$$

hence, *four times the arc whose tangent is* $\dfrac{1}{5}$, *exceeds the arc of 45°*

by an arc whose tangent is $\dfrac{1}{239}$.

SECTION VII.

Classification of Curves.

101. CURVES may be divided into two general classes:
1st. Those whose equations are purely algebraic; and
2dly. Those whose equations involve transcendental quantities.

Those of the first class, are called Algebraic curves, and those of the second, *Transcendental curves*.

The properties of the Algebraic curves have been already examined; it therefore only remains to explain the proper-t.es of the Transcendental curves.

Logarithmic Curve.

102. A logarithmic curve, is a curve in which one of the co-ordinates, of any point, is the logarithm of the other. The co-ordinate axis to which the lines denoting the logarithms are parallel, is called the *axis of logarithms*, and the other, the *axis of numbers*.

If we suppose Y to be the axis of logarithms, then X will be the axis of numbers, and the equation of the curve will be,

$$y = \log x.$$

General Properties.

103. Let A be the origin of a system of rectangular co-ordinates, X the axis of numbers, and Y the axis of logarithms.

If we designate the base of a system of logarithms by a, we shall have,*

$$a^y = x,$$

in which y is the logarithm of x.

If we change the value of the base a, to a', we shall have,

$$a'^y = x,$$

in which y is the logarithm of x, to the base a'. It is plain, that the same value of x, in the two equations, will give different values of y, and hence : *Each system of logarithms will give a different logarithmic curve.*

If we make $y = 0$, we shall have,† $x = 1$; and since this relation is independent of the base of the system of logarithms, it follows, that: *Every logarithmic curve will intersect the axis of numbers at a distance from the origin equal to 1.*

This abscissa is denoted by the line AE.

We may find points of the curve from the general equation,

$$a^y = x,$$

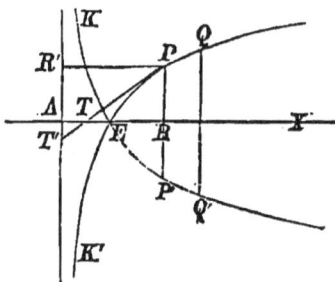

* Bourdon, Art. **227**. University, Art. **183**.
† Bourdon, Art. **235**. University, Art. **186**.

even without the aid of a table of logarithms.   For, if
we make,

$$y = 0, \qquad y = \frac{1}{2}, \qquad y = \frac{3}{2}, \qquad y = \frac{1}{4}, \quad \&c.,$$

we shall find, for the corresponding values of x,

$$x = 1, \qquad x = \sqrt{a}, \qquad x = a\sqrt{a}, \qquad x = \sqrt[4]{a}, \quad \&c.$$

If we make $a = 10$, the curve will correspond to the
common system of logarithms; and if we suppose
$a = 2.7182818...$, to the Naperian system.   Both curves
pass through the point E.

Base > 1.

104. If we suppose the base of the system of logarithms to be greater than 1, the logarithms of all numbers
less than 1 will be negative;* therefore, the values of y,
corresponding to all abscissas between the limits of $x = 0$,
and $x = AE = 1$, will be negative; hence, these ordinates are laid off below the axis of X.   When $x = 0$,
$y = -\infty$.   Therefore, when the base is greater than 1, the
corresponding curve is $QPEK'$.   The curve cannot extend
to the left of the axis of Y, since negative numbers have
no real logarithms.†

Base < 1.

105. If the base of the system is less than 1, the logarithms of all numbers greater than 1 are negative; and
of all numbers less than 1, positive.   Under this supposition, the curve assumes the position $Q'P'EK$.   The parts

---

* Bourdon, Art. **235.**   University, Art. **186.**
† Bourdon, Art. **235.**   University, Art. **186**

of the curves EPQ, $EP'Q'$, are concave towards the axis of abscissas; the parts EK, EK', are convex; and both curves, throughout their whole extent, are convex towards the axis of Y.

Asymptote.

106. Let us resume the equation of the curve,

$$y = \log x.$$

If we denote the modulus of a system of logarithms by M, and differentiate, we have (Art. **88**),

$$dy = \frac{dx}{x}M; \quad \text{or,} \quad \frac{dy}{dx} = \frac{M}{x}.$$

But, $\frac{dy}{dx}$ denotes the tangent of the angle which the tangent line makes with the axis of abscissas; hence, the tangent will be parallel to the axis of abscissas when $x = \infty$, and perpendicular to it, when $x = 0$.

But, when $x = 0$, $y = -\infty$; hence, the axis of ordinates is an asymptote to the curve. The tangent which is parallel to the axis of X, is not an asymptote; for, when $x = \infty$, we also have, $y = \infty$ (Art. **50**).

Sub-tangent.

107. The most remarkable property of this curve, is the value of its sub-tangent $T'R'$, estimated on the axis of logarithms. We have found, for the sub-tangent, on the axis of X (Art. **45**),

$$TR = \frac{dx}{dy}y;$$

and by simply changing the axis, we have,

$$T'R' = \frac{dy}{dx} x = M \text{ (Art. 106);} \quad \text{hence,}$$

The sub-tangent, taken on the axis of logarithms, is equal to the modulus of the system from which the curve is constructed. In the Naperian system, $M = 1$; hence, the sub-tangent is equal to 1, equal to AE. In the common system, it is denoted by the number, .434284482 . . .

The Cycloid.

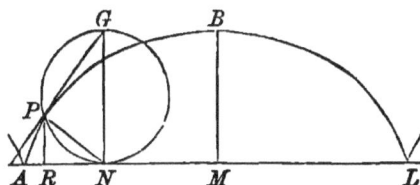

108. If a circle NPG be rolled along a straight line, .AL, any point of the circumference, as P, will describe a curve, which is called a *cycloid*. The circle NPG is called the *generating circle*, and P, the *generating point*.

Since each succeeding revolution of the generating circle will describe an equal curve, it will only be necessary to examine the properties of the curve $APBL$, described in one revolution. We shall, therefore, refer only to this part, when speaking of the cycloid.

If we suppose the point P to be on the line AL, at A, it will be found at some point, as L, after all the points of the circumference shall have been brought in contact with the line AL. The line AL will be equal to the circumference of the generating circle, and is called the *base of the cycloid.* The line BM, drawn perpendicular to the base, at the middle point, is called the *axis of the cycloid*, and is equal to the diameter of the generating circle.

Transcendental Equation of the Cycloid.

109. Let CN be the radius of the generating circle. Assume any point, as A, for the origin of co-ordinates. Let us suppose that when the generating point has described any arc of the cycloid, as AP, that the point in which the circle touches the base has reached the point N.

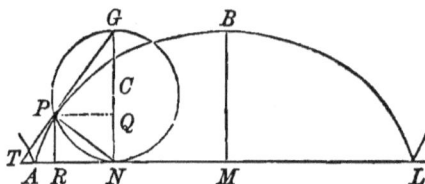

Through N, draw the diameter NG, of the generating circle : it will be perpendicular to the base AL. Through P, draw PR perpendicular to the base, and PQ parallel to it. Then, $PR = NQ$ will be the versed sine, and PQ the sine of the arc NP to the radius CN. Put,

$$CN = r, \qquad AR = x, \qquad PR = NQ = y;$$

we shall then have,

$$PQ = \sqrt{2ry - y^2}, \quad x = AN - RN = \text{arc } NP - PQ;$$

hence, the transcendental equation is,

$$x = \text{ver-sin}^{-1} y - \sqrt{2ry - y^2}.$$

Differential Equation.

110. The properties of the cycloid are most easily deduced from its differential equation. This is found by differentiating both members of the transcendental equation. We have (Art. **97**),

$$d(\text{ver-sin}^{-1}y) = \frac{rdy}{\sqrt{2ry - y^2}}; \quad \text{and}$$

$$d\left(-\sqrt{2ry - y^2}\right) = -\frac{rdy - ydy}{\sqrt{2ry - y^2}}; \quad \text{hence,}$$

$$dx = \frac{rdy}{\sqrt{2ry - y^2}} - \frac{rdy - ydy}{\sqrt{2ry - y^2}}; \quad \text{or,} \quad dx = \frac{ydy}{\sqrt{2ry - y^2}};$$

which is the differential equation of the cycloid.

Sub-Tangent, Tangent, Sub-Normal, Normal.

111. If we substitute in the general equations of Arts. **45, 46, 47,** and **48,** the value of $\frac{dy}{dx}$, found in the differential equation of the cycloid, we shall obtain the values of the sub-tangent, tangent, normal, and sub-normal.

$$TR = \frac{y^2}{\sqrt{2ry - y^2}} = \text{sub-tangent};$$

$$TP = \frac{y\sqrt{2ry}}{\sqrt{2ry - y^2}} = \text{tangent};$$

$$PN = \sqrt{2ry} = \text{normal};$$

$$RN = \sqrt{2ry - y^2} = \text{sub-normal}.$$

These values are easily constructed, from their connection with the parts of the generating circle.

The sub-normal RN, for example, is equal to PQ of the generating circle, since each is equal to $\sqrt{2ry - y^2}$; hence, the normal PN, and the diameter GN, intersect the base of the cycloid at the same point. Now, since the tangent to the cycloid at the point P must be perpendicular to the normal, it will coincide with the chord PG of the generating circle.

If, therefore, it be required to draw a normal, or a tangent, to the cycloid, at any point, as P, draw any line, as ng, perpendicular to the base AL, and make it equal to the diameter of the generating circle. On ng, describe a semi-circumference, and through P draw a parallel to the base of the cycloid. Through p, where the parallel cuts the semi-circumference, draw the supplementary chords pn, pg, and then draw through P the parallels PN, PG; and PN will be a normal, and PG a tangent to the cycloid at the point P.

Position of Tangent.

112. The differential equation of the curve,

$$dx = \frac{ydy}{\sqrt{2ry - y^2}},$$

may be put under the form,

$$\frac{dy}{dx} = \frac{\sqrt{2ry - y^2}}{y} = \sqrt{\frac{2r}{y} - 1}.$$

If we make $y = 0$, we shall have,

$$\frac{dy}{dx} = \infty;$$

and if we make $y = 2r$, we shall have,

$$\frac{dy}{dx} = 0;$$

hence, the tangent lines drawn to the cycloid at the points where the curve meets the base, are perpendicular to the base; and the tangent drawn through the extremity of the greatest ordinate, is parallel to the base.

Curve Concave.

113. If we differentiate the equation,

$$dx = \frac{ydy}{\sqrt{2ry - y^2}},$$

regarding dx as constant, we obtain,

$$0 = \sqrt{2ry - y^2}\,(yd^2y + dy^2) - \frac{ydy(rdy - ydy)}{\sqrt{2ry - y^2}};$$

or, by reducing and dividing by y,

$$0 = (2ry - y^2)d^2y + rdy^2,$$

whence we obtain,

$$d^2y = - \frac{rdy^2}{2ry - y^2};$$

and hence, the curve is concave towards the axis of abscissas (Art. **73**).

Area of the Cycloid.

114. The area of the cycloid may be found in a very simple manner, by constructing the rectangle $AFBM$, and considering the portion AFB.

If we regard F as an origin of co-ordinates, FB as a line of abscissas, and take any ordinate, as,

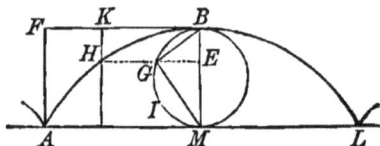

$$KH = z = 2r - y,$$

we shall have, $d(AHKF) = zdx.$

But, $zdx = \dfrac{(2r - y)ydy}{\sqrt{2ry - y^2}} = dy\sqrt{2ry - y^2};$

whence, $AHKF = \int dy\sqrt{2ry - y^2} + C.$

But this integral expresses the area of the segment of a circle, whose radius is r, and versed-sine y (Art. **99**), that is, of the segment $MIGE$. If now, we estimate the area of the segment from M, where $y = 0$, and the area $AFKH$ from AF, in which case the area $AFKH = 0$, for $y = 0$, we shall have,

$$AFKH = MIGE;$$

and taking the integral between the limits $y = 0$ and $y = 2r$, we have,

$$AFB = \text{semi-circle } MIGB,$$

and consequently,

$$\text{area } AHBM = AFBM - MIGB.$$

But the base of the rectangle $AFBM$ is equal to the semi-circumference of the generating circle, and the altitude is equal to the diameter; hence, its area is equal to four times the area of the semi-circle $MIGB$; therefore,

$$\text{area } AHBM = 3MIGB; \text{ hence,}$$

The area $AHBL$ is equal to three times the area of the generating circle.

Surface described by the Cycloid.

115. To find the surface described by the arc of a cycloid when revolved about its base.

The differential equation of the cycloid is,

$$dx = \frac{ydy}{\sqrt{2ry - y^2}}.$$

Substituting this value of dx in the differential equation of the surface (Art. **62**), it becomes,

$$ds = \frac{2\pi\sqrt{2r}\,y^{\frac{3}{2}}dy}{\sqrt{2ry - y^2}}.$$

Applying Formula (E), (Art. **170**), we have,

$$s = 2\pi\sqrt{2r}\left[-\frac{2}{3}y^{\frac{1}{2}}\sqrt{2ry - y^2} + \frac{4}{3}r\int\frac{y^{\frac{1}{2}}dy}{\sqrt{2ry - y^2}}\right].$$

But,

$$\int\frac{y^{\frac{1}{2}}dy}{\sqrt{2ry - y^2}} = \int\frac{dy}{\sqrt{2r - y}} = \int dy(2r - y)^{-\frac{1}{2}} = -2(2r - y)^{\frac{1}{2}};$$

hence,

$$s = 2\pi\sqrt{2r}\left[-\frac{2}{3}y^{\frac{1}{2}}\sqrt{2ry - y^2} - \frac{8}{3}r(2r - y)^{\frac{1}{2}}\right] + C.$$

If we estimate the surface from the plane passing through the centre, we have $C = 0$, since at this point $s = 0$, and $y = 2r$. If we then integrate between the limits $y = 2r$, and $y = 0$, we have,

$$s = \frac{1}{2}\text{ surface} = -\frac{32}{3}\pi r^2; \quad\text{hence,}$$

$$s = \text{surface} = -\frac{64}{3}\pi r^2,$$

that is, the surface described by the cycloid, when it is revolved around the base, is equal to 64 thirds of the generating circle.

The minus sign should appear before the integral, since the surface is a decreasing function of the variable y (Art. **19**).

Volume generated by the area of the Cycloid.

116. If a cycloid be revolved about its base, it is required to find the measure of the volume which the area will generate.

The differential equation of the cycloid is,

$$dx = \frac{y\,dy}{\sqrt{2r - y^2}}.$$

If **we** denote the volume by V, we have (Art. **69**),

$$dV = \frac{\pi y^3 dy}{\sqrt{2ry - y^2}}.$$

If we apply Formula (E) (Art. 170), we shall find, after three reductions, that the integral will depend on that of

$$\frac{dy}{\sqrt{2ry - y^2}}.$$

But the integral of this expression is the arc whose versed sine is $\frac{y}{r}$ (Art. 99). Making the substitutions and reductions, we find the volume equal to five-eighths of the circumscribing cylinder.

Spirals.

117. A *Spiral*, or *Polar Line*, is a curve described by a point which moves along a right line, according to any law whatever, the line having at the same time a uniform angular motion.

Let ABC be a straight line which is to be turned uniformly around the point A. When the motion of the line begins, let us suppose a point to move from A along the line, in the direction ABC. When the line takes the position ADE, the point will have moved along it, to some point, as D, and will have described the arc AaD, of the spiral. When the line takes the position $AD'E'$, the point will have described the curve $AaDD'$, and when the line shall have completed an entire revolution, the point will have described the curve $AaDD'B$.

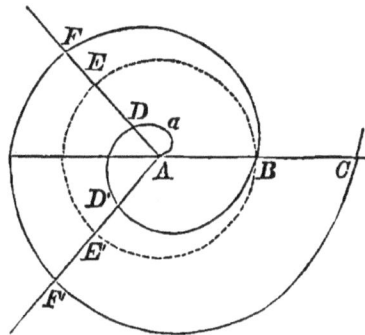

If the revolutions of the radius-vector be continued, the

generating point will describe an indefinite spiral. The point A, about which the right line revolves, is called the *pole ;* the distances AD, AD', AB, are called *radius-vectors* or *radii-vectores ;* and the parts $AaDD'B$, $BFF''C$, described in each revolution, are called *spires.*

If, with the pole as a centre, and AB, the distance passed over by the generating point in the direction of the radius-vector, during the first revolution, as a radius, we describe the circumference BEE', the angular motion of the radius-vector about the pole A, may be measured by the arcs of this circle, estimated from B.

If we designate the radius-vector by u, and the measuring arc, estimated from B, by t, the relation between u and t, may be expressed by the equation,

$$u = f(t), \text{ or } u = at^n,$$

in which n depends on the *law* according to which the generating point moves along the radius-vector, and a on the relation which exists between a *given* value of u, and the corresponding value of t.

General Properties.

118. When n is positive, the spirals represented by the equation,

$$u = at^n,$$

will pass through the pole A. For, if we make $t = 0$, we shall have, $u = 0$.

But if n is negative, the equation will become,

$$u = at^{-n}; \quad \text{or,} \quad u = \frac{a}{t^n},$$

from which we shall have,

for, $t = 0$, $u = \infty$,

and for, $t = \infty$, $u = 0$;

hence, in this class of spirals, the first position of the generating point is at an *infinite distance* from the pole: the point will then approach the pole as the radius-vector revolves, and will only reach it after an *infinite number* of revolutions.

Spiral of Archimedes.

119. If we make $n = 1$, the equation of the spiral becomes,

$$u = at.$$

If we designate two different radii-vectores by u' and u'', and the corresponding arcs by t' and t'', we shall have,

$$u' = at', \quad \text{and} \quad u'' = at'',$$

and consequently,

$$u' : u'' :: t' : t''; \quad \text{that is,}$$

The radii-vectores are proportional to the measuring arcs, estimated from the initial point.

This spiral is called the spiral of Archimedes.

If we denote by 1, the distance which the generating point moves along the radius-vector, during one revolution, the equation,

$$u = at,$$

will become,

$$1 = at; \quad \text{or,} \quad 1 \times \frac{1}{a} = t.$$

But since t is the circumference of a circle whose radius is 1, we shall have,

$$\frac{1}{a} = 2\pi, \quad \text{and consequently,} \quad a = \frac{1}{2\pi}.$$

Parabolic Spiral.

120. If we make $n = \frac{1}{2}$, and $a = \sqrt{2p}$, we have, for the general equation,

$$u = \sqrt{2p} \times t^{\frac{1}{2}}; \quad \text{or,} \quad u^2 = 2pt,$$

which is the equation of the parabolic spiral.

If $t = 0$, $u = 0$; hence, this spiral passes through the pole.

Hyperbolic Spiral.

121. If we make $n = -1$, the general equation of spirals becomes,

$$u = at^{-1}; \quad \text{or,} \quad ut = a.$$

This spiral is called the *hyperbolic spiral*, because of the analogy which its equation bears to that of the hyperbola, when referred to its asymptotes.

If, in this equation, we make, successively,

$$t = 1, \quad t = \frac{1}{2}, \quad t = \frac{1}{3}, \quad t = \frac{1}{4}, \quad \&c.,$$

we shall have the corresponding values,

$$u = a, \quad u = 2a, \quad u = 3a, \quad u = 4a, \quad \&c.$$

Logarithmic Spiral.

122. Since the relation between u and t is entirely arbitrary, we may, if we please, make,

$$t = \log u.$$

The spiral described by the extremity of the radius-vector, under this supposition, is called the *logarithmic spiral*.

Direction of the measuring arc.

123. The arc, which measures the angular motion of the radius-vector, has been estimated from right to left, and the value of t regarded as positive. If we revolve the radius-vector in a contrary direction, the measuring arc will be estimated from left to right, the sign of t will be changed to negative, and a similar spiral will be described.

Sub-tangent in Polar Curves.

124. The SUB-TANGENT, in spirals, or in *any curve*, referred to polar co-ordinates, is the projection of the tangent on a line drawn through the pole, and perpendicular to the radius-vector passing through the point of contact.

Let A be the pole, $AN = 1$, the radius of the measuring arc, P any point of the curve, TP a tangent at P, and AT;

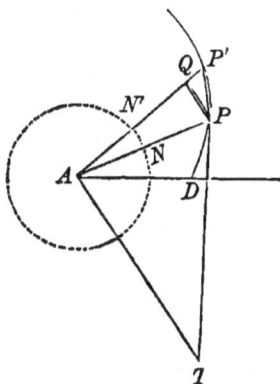

perpendicular to AP, the sub-tangent. Let AP' be a radius-vector, consecutive with AP, and PQ, an arc described from the centre A.

Then, $NN' = dt$, and $QP = du$, and, since PQ is parallel to NN', we have, $PQ = udt$. But the arc PQ coincides with its chord (Art. **43**), and since Q is a right angle, the triangles PQP' and TAP are similar; hence,

$$AT : AP :: PQ : QP'; \text{ therefore,}$$

Sub-tangent $\quad AT : u :: udt : du.$

Whence, $\quad$ Sub-tangent $AT = \dfrac{u^2 dt}{du} = \dfrac{a}{n}t^{n+1}.$

125. In the spiral of Archimedes, we have,

$$n = 1, \quad \text{and} \quad a = \frac{1}{2\pi};$$

hence, $\qquad AT = \dfrac{t^2}{2\pi}.$

If we make $t = 2\pi$, circumference of the measuring circle, we shall have,

$AT = 2\pi$, circumference of the measuring circle.

After m revolutions, we shall have

$$t = 2m\pi,$$

and consequently,

$$AT = 2m^2\pi = m.2m\pi; \text{ that is,}$$

The sub-tangent, after m revolutions, is equal to m times

the circumference of the circle whose radius is the radius-vector. This property was discovered by Archimedes.

126. In the hyperbolic spiral, $n = -1$, and the value of the sub-tangent becomes

$$AT = -a; \text{ that is,}$$

The sub-tangent is constant in the hyperbolic spiral.

Angle of the Tangent and Radius-Vector.

127. We see that,

$$\frac{AT}{AP} = \frac{udt}{du},$$

denotes the tangent of the angle which the tangent line makes with the radius-vector.

In the logarithmic spiral, of which the equation is

$$t = \log u,$$

we have,
$$dt = \frac{du}{u}M;$$

hence,
$$\frac{AT}{AP} = \frac{udt}{du} = M; \text{ that is,}$$

In the logarithmic spiral, the angle formed by the tangent and the radius-vector passing through the point of contact, is constant; and the tangent of the angle is equal to the modulus of the system of logarithms.

If t is the Naperian logarithm of u, M is 1 (Art. **89**), and the angle will be equal to 45°.

Value of the Tangent.

128. The value of the tangent, in a curve referred to polar co-ordinates, is,

$$PT = \sqrt{\overline{AP}^2 + \overline{AT}^2} = u\sqrt{1 + \frac{u^2 dt^2}{du^2}}.$$

Differential of the Arc.

129. To find the differential of the arc, which we denote by z, we have,

$$PP' = \sqrt{\overline{QP'}^2 + \overline{QP}^2};$$

or, by substituting for PP', QP', and PQ, their values, when P and P' are consecutive, we have,

$$dz = \sqrt{du^2 + u^2 dt^2}.$$

Differential of the Area.

130. The differential of the area ADP, when referred to polar co-ordinates, is not an elementary rectangle, as when referred to rectangular axes, but is the elementary sector APP'. The area of this triangle is equal to $\dfrac{AP' \times PQ}{2}$. If we denote the differential by ds, we have,

$$ds = \frac{AP' \times QP}{2} = \frac{(u + du)udt}{2};$$

or, omitting the infinitely small quantity of the second order, $ududt$ (Art. **20**),

$$ds = \frac{u^2 dt}{2},$$

which is the differential of the area of any segment of a polar line.

Areas of Spirals.

131. If we denote by s, the area described by the radius-vector, we have (Art. **130**),

$$ds = \frac{u^2 dt}{2};$$

and placing for u its value, at^n (Art. **117**),

$$ds = \frac{a^2 t^{2n} dt}{2}, \quad \text{and} \quad s = \frac{a^2 t^{2n+1}}{4n+2} + C.$$

If n is positive, $C = 0$, since the area is 0, when $t = 0$. After one revolution of the radius-vector, $t = 2\pi$, and we have,

$$s = \frac{a^2 (2\pi)^{2n+1}}{4n+2},$$

which is the area included within the first spire.

132. In the spiral of Archimedes, (Art. **119**),

$$a = \frac{1}{2\pi}, \quad \text{and} \quad n = 1;$$

hence, for this spiral we have,

$$s = \frac{t^3}{24\pi^2},$$

which becomes $\frac{\pi}{3}$, after one revolution of the radius-vector; the unit of the number $\frac{\pi}{3}$, being a square whose side is 1. Hence, *the area included by the first spire, is equal to one-third of the area of the circle whose radius is the radius-vector, after the first revolution.*

In the second revolution, the radius-vector describes a second time, the area described in the first revolution; and in any succeeding revolution, it will pass over, or re-describe, all the area before generated. Hence, to find the area, at the end of the mth revolution, we must integrate between the limits,

$$t = (m - 1)2\pi, \quad \text{and} \quad t = m.2\pi,$$

which gives,

$$s = \frac{m^3 - (m - 1)^3}{3}\pi.$$

If it be required to find the area between any two spires, as between the mth and the $(m + 1)$th, we have for the whole area to the $(m + 1)$th spire,

$$\frac{(m + 1)^3 - m^3}{3}\pi;$$

and subtracting the area to the mth spire, gives,

$$s = \frac{(m + 1)^3 - 2m^3 + (m - 1)^3}{3}\pi = 2m\pi,$$

for the area between the mth and $(m + 1)$th spires.

If we make $m = 1$, we shall have the area between the first and second spires equal to 2π; hence, *the area between the mth and $(m + 1)$th spires, is equal to m times the area between the first and second.*

133. In the hyperbolic spiral, $n = -1$, and we have,

$$ds = \frac{a^2 t^{-2}}{2} dt, \quad \text{and} \quad s = -\frac{a^2}{2t}.$$

The area s will be infinite, when $t = 0$, but we can

find the area included between any two radii-vectores b and c, by integrating between the limits $t = b$ and $t = c$, which will give,

$$s = \frac{a^2}{2}\left(\frac{1}{b} - \frac{1}{c}\right).$$

134. In the logarithmic spiral, $t = lu$; hence,

$$dt = \frac{du}{u},$$

and, $\quad \dfrac{u^2 dt}{2} = \dfrac{u\,du}{2};$

hence, $\quad s = \displaystyle\int \frac{u\,du}{2} = \frac{u^2}{4} + C;$

and by considering the area $s = 0$, when $u = 0$, we have $C = 0$, and

$$s = \frac{u^2}{4}.$$

CURVATURE.

135. THE CURVATURE of a plane curve, at any point, is the *departure* from the tangent drawn to the curve at that point. This departure is measured by the distance which a point, moving on the curve, departs from the tangent in passing over a unit of length, denoted by the differential of the arc. In the same circle, or in equal circles, the departure from a tangent, at any point, is always the same; hence, the curvature of a circle, at all points, is constant.

Curvature of a circle is inversely as the radius.

136. Let C and C' be the centres of two unequal circles, having a common tangent at P. If we suppose the arcs to be the independent variables, we can denote the differential of one arc by Pb, and the differential of the other, by an equal arc Pa. Then, having drawn bB and aD, and the sines, bb', aa', and recollecting that each arc is equal to its corresponding chord, (Art. **43**), we have, by denoting the radii by r and r',*

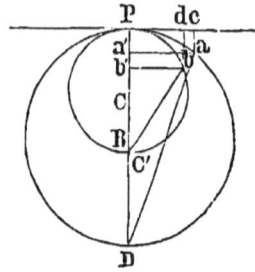

$$\overline{Pb}^2 = 2r.Pb', \quad \text{and} \quad \overline{Pa}^2 = 2r'.Pa';$$

since the arcs are equal, and $Pb' = db$, and $Pa' = ca$,

$$2r.db = 2r'.ca; \quad \text{hence,}$$

$$db \;:\; ac \;::\; \frac{1}{r} \;:\; \frac{1}{r'}; \quad \text{that is,}$$

The curvature of a circle varies inversely as its radius; hence, *the reciprocal of the radius of a circle may be assumed as the measure of its curvature.*

Orders of Contact.

137. If two plane curves have one point in common, there is one set of co-ordinates (which may be denoted by x'', y''), that will satisfy the equations of both curves. If the curves have a second point in common, *consecutive with the first*, they will have a common tangent, at the common point, and the first differential coefficients will also be equal (Art. **43**); this is called, *a contact of the first order*, If the curves have a third point in common, consecutive with the second, the second differential coefficients will be

---

* Legendre, Bk. IV. P. 23.

equal (Art. **73**) ; this is called, *a contact of the second order.*

Generally, two curves have a contact of the nth order, when they have a common point, and the first n successive differential coefficients of the common ordinate, equal to each other.

Osculatory Curves.

138. An Osculatrix, is a curve which has a highei order of contact with a given curve, at a given point, than any other curve of the same kind. The osculatory circle is by far the most important of all the osculatrices; for it is this circle which measures the curvature of all plane curves.

Osculatory Circle.

139. The general equation of a circle, referred to rectangular co-ordinates (Bk. II., Art. **5**), is,

$$(x - \alpha)^2 + (y - \beta)^2 = R^2 \quad . \quad . \quad . \quad (1.)$$

in which α and β are the co-ordinates of the centre, and x and y the co-ordinates of any point of the curve.

If we regard α, β, and R, as constants, and differentiate the equation twice, and then find the differential coefficients of the first and second order, we have,

$$\frac{dy}{dx} = - \frac{x - \alpha}{y - \beta} \quad . \quad . \quad . \quad . \quad (2.)$$

and,
$$\frac{d^2y}{dx^2} = - \frac{1 + \dfrac{dy^2}{dx^2}}{y - \beta} \quad . \quad . \quad . \quad (3.)$$

In Equation (1) there are three arbitrary constants, α, β, and R; and values may be assigned to these, at pleasure, so as to cause the circle to fullfil three conditions, and three only.

If we have any plane curve whose equation is of the form,

$$y = f(x) \, ;$$

and find, from this equation, the first and second differential coefficients, for any point whose co-ordinates are x'', y'', we may then attribute such values to α, β, and R, as shall make,

$$x = x'', \qquad y = y''; \qquad \text{also,}$$

$$\frac{dy}{dx} = \frac{dy''}{dx''}, \qquad \text{and} \qquad \frac{d^2y}{dx^2} = \frac{d^2y''}{dx''^2}.$$

As no further general relations can be established between the differential coefficients of the circle and curve, this circle will be osculatory to the curve at the point whose co-ordinates are x'', y'' (Art. **138**). Since the co-ordinates of a point, and the differential coefficients of the first and second order, determine three consecutive points (Art. **73**), it follows that, *the osculatory circle passes through three consecutive points of the curve, at the point of osculation.*

Limit of the Orders of Contact.

140. It is seen that the highest order of contact which a circle can have with any curve, is denoted by the number of arbitrary constants which enters into its equation, less 1; and the same is true for any other osculatrix.

Although it is impossible to *assign* a higher order of contact, to a circle, than the *second*, yet, at the vertices of the transverse and conjugate axes of the conic sections, the conditions which make the circle osculatory, also make the third differential coefficient zero, and hence give a contact of the third order. In general, when the order of contact is *even*, and the curve symmetrical with the normal at the point of osculation, the conditions imposed will give a contact of the next higher order.

Radius of Curvature.

141. If we find the value of R from Equations (1), (2), and (3), we have,

$$R = \pm \frac{(dx^2 + dy^2)^{\frac{3}{2}}}{dx d^2 y} \quad \ldots \ldots (4.)$$

which is the general value for the radius of the osculatory circle.

If we denote the arc by z, we have (Art. **52**),

$$dz = \sqrt{dx^2 + dy^2};$$

whence,
$$R = \pm \frac{dz^3}{dx d^2 y} \quad \ldots \ldots (5.)$$

Measure of Curvature.

142. The curvature of a curve, at any point, is measured by the curvature of the osculatory circle at that point; hence, it is the reciprocal of the radius (Art. **136**).

If we assume two points, P and P', either on the

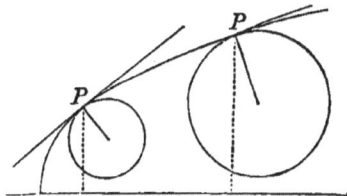

same, or on different curves, and find the radii r and r' of the circles which are osculatory at these points, then,

$$\text{curvature at } P : \text{curvature at } P' :: \frac{1}{r} : \frac{1}{r'}.$$

143. To find the radius of curvature, at any point of a plane curve, whose equation is of the form,

$$y = f(x).$$

Differentiate the equation twice, and substitute the values of the first and second differentials in Equation (4); the resulting equation will indicate the value of R for that point.

If we use the $+$ sign, when the curve is convex toward the axis of abscissas, and the $-$ sign when it is concave, the essential sign of R will be positive, when R is an increasing function of x.

Radius of Curvature for Lines of the Second Order.

144. The general equation for lines of the second order (Bk. V, Art. **42**), is,

$$y^2 = mx + nx^2,$$

which gives, by differentiation,

$$dy = \frac{(m + 2nx)dx}{2y}, \quad dx^2 + dy^2 = \frac{[4y^2 + (m + 2nx)^2]dx^2}{4y^2}$$

$$d^2y = \frac{2ny\,dx^2 - (m + 2nx)dx\,dy}{2y^2} = \frac{[4ny^2 - (m + 2nx)^2]dx^2}{4y^3}$$

Substituting these values in the equation,

$$R = - \frac{(dx^2 + dy^2)^{\frac{3}{2}}}{dx d^2 y},$$

we obtain, $\qquad R = \frac{[4(mx + nx^2) + (m + 2nx)^2]^{\frac{3}{2}}}{2m^2};$

which is the radius of curvature in lines of the second order, for any abscissa x.

145. If we make $x = 0$, we have,

$$R = \frac{1}{2}m = \frac{B^2}{A};$$

that is, in lines of the second order, *the radius of curvature at the vertex of the transverse axis is equal to half the parameter of that axis.*

146. If it is required to find the value of the radius of curvature at the vertex of the conjugate axis of an ellipse, we make (Bk. V, Art. **42**),

$$m = \frac{2B^2}{A}, \qquad n = - \frac{B^2}{A^2}, \qquad \text{and} \qquad x = A,$$

which gives, after reducing,

$$R = \frac{A^2}{B}; \quad \text{hence,}$$

The radius of curvature at the vertex of the conjugate axis of an ellipse is equal to half the parameter of that axis.

147. In the case of the parabola, in which $n = 0$, the general value of the radius of curvature becomes,

$$R = \frac{(m^2 + 4mx)^{\frac{3}{2}}}{2m^2}.$$

If we make $x = 0$, we shall have the radius of curvature at the vertex, equal to $\dfrac{m}{2}$, or *one-half the parameter.*

148. If we compare the value of the radius of curvature (Art. **144**), with that of the normal line found in Art. **49**, we shall have,

$$R = \frac{(\text{normal})^3}{\frac{1}{4}\,m^2}; \text{ that is,}$$

The radius of curvature, at any point, is equal to the cube of the normal divided by half the parameter squared; and hence, the radii of curvature, at different points of the same curve, are to each other as the cubes of the corresponding normals; and the curvature is proportional to the reciprocals of those cubes.

Evolute Curves.

149. An EVOLUTE curve is the locus of the centres of all the circles which are osculatory to a given curve. The
given curve is called the INVO-
LUTE.

If at different points, P, P',
P'', &c., of an involute, or given
curve, normals, PC, $P'C'$, &c.,
be drawn, and distances laid off
on them, on the concave side of
the arc, each equal to the radius of curvature at the
point, then the curve drawn through the extremities C,
C', C'', &c., of these radii of curvature, is the evolute
curve.

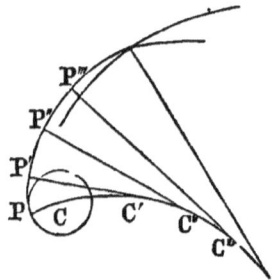

A normal to the Involute is tangent to the Evolute.

150. Resuming the consideration of the first three equations of Art. **139**, and changing slightly the forms of (2) and (3), we have,

$$(x - a)^2 + (y - \beta)^2 = R^2 \quad . \quad . \quad . \quad (1.)$$

$$(x - a)dx + (y - \beta)dy = 0 \quad . \quad . \quad . \quad (2.)$$

$$dx^2 + dy^2 + (y - \beta)d^2y = 0 \quad . \quad . \quad . \quad (3.)$$

Equations (2) and (3) were derived from Equation (1), under the supposition that a, β, and R, were arbitrary constants, and of such values as to cause the circle to be osculatory to a given curve, at a given point.

If now, we suppose the osculatory circle to move along the involute, continuing osculatory to it, the five quantities, R, a, β, y, dy, will all be functions of the independent variable x, and a and β will be the co-ordinates of the evolute curve.

If we differentiate Equations (1) and (2) under this hypothesis, we have,

$$(x - a)dx + (y - \beta)dy - (x - a)da - (y - \beta)d\beta = RdR,$$

$$dx^2 + dy^2 + (y - \beta)d^2y - dadx - d\beta dy = 0.$$

Combining the first with Equation (2), and the second with (3), we obtain,

$$- (y - \beta) d\beta - (x - a)da = RdR, \quad . \quad (4.)$$

$$- dadx - d\beta dy = 0 \quad . \quad . \quad . \quad (5.)$$

From the last equation we have,

$$\frac{d\beta}{d\alpha} = -\frac{dx}{dy} \quad \dots \dots \quad (6.)$$

But Equation (2) may be placed under the form,

$$y - \beta = -\frac{dx}{dy}(x - \alpha), \quad \text{or,} \quad \beta - y = -\frac{dx}{dy}(\alpha - x) \quad . \ (7.)$$

Substituting for $-\dfrac{dx}{dy}$, its value $\dfrac{d\beta}{d\alpha}$, we have,

$$y - \beta = \frac{d\beta}{d\alpha}(x - \alpha) \quad \dots \dots \quad (8.)$$

Since Equations (7) and (8) are the same under different forms, they represent one and the same line.

Equation (7) is the equation of a normal to the involute at a point whose co-ordinates are x and y, and passes through any point whose co-ordinates are α and β (Art. **44**). Equation (8) is the equation of a tangent to the evolute at a point whose co-ordinates are α and β, and passes through any point whose co-ordinates are x and y (Art. **43**); therefore,

The radius of curvature which is normal to the involute is tangent to the evolute.

Evolute and radius of curvature increase or decrease by the same quantity.

151. Combining Equations (2) and (6), we have,

$$x - \alpha = \frac{d\alpha}{d\beta}(y - \beta) \quad \dots \dots \quad (9.)$$

Substituting this value of $x - \alpha$ in Equation (1), we have, after reduction,

$$(y - \beta)^2 \left(\frac{d\alpha^2 + d_i\beta^2}{d_i\beta^2} \right) = R^2 \quad . \quad . \quad (10.)$$

Substituting the same value in Equation (4), reducing, and squaring both members, we obtain,

$$(y - \beta)^2 \frac{(d\alpha^2 + d_i\beta^2)^2}{d_i\beta^2} = R^2(dR)^2 \quad . \quad (11.)$$

Dividing (11) by (10), member by member, and taking the root,

$$\sqrt{d\alpha^2 + d_i\beta^2} = dR \quad . \quad . \quad . \quad . \quad (12.)$$

But since α and β are the co-ordinates of the evolute, if we denote this curve by z, we shall have (Art. **52**),

$$dR = dz, \quad dR - dz = 0, \quad d(R - z) = 0;$$

whence,   · $R - z =$ a constant (Art. **17**);

which, if we denote by c, gives,

$$R = z + c \quad . \quad . \quad . \quad . \quad (13.)$$

Since the difference between R and z is constant, it follows that any change in one, will produce a *corresponding* and *equal change* in the other.

If we draw any two radii of curvature, as PC, $P'C'$, and denote them by R and R', and the corresponding arcs of the evolute by z and z', we have,

$$R = z + c, \quad \text{and} \quad R' = z' + c;$$

whence, $R' - R = z' - z;$ that is,

The difference between any two radii of curvature is

equal to the arc of the evolute intercepted between their extremities.

If we make $z = 0$, and denote the corresponding value of R by r, we have,

$$r = 0 + c = c; \text{ hence,}$$

The constant c, is equal to the radius of curvature passing through the origin of arc of the evolute.

If we suppose C to be the origin of arc of the evolute, then, $CP = r = c$; and any radius of curvature, as $C'P'$, will be equal in length to the line $C'CP$. If then the evolute be developed, or unrolled, as it were, about the movable centre of the osculatory circle, the other extremity of the radius of curvature will describe the involute curve.

Evolute of the Cycloid.

152. Let us resume the equation for the radius of curvature (Art. **141**),

$$R = -\frac{(dx^2 + dy^2)^{\frac{3}{2}}}{dx d^2 y} \quad \dots \quad (1.)$$

If, in this equation, we substitute the value of dx and $d^2 y$, found in Art. **113**, we have,

$$R = 2^{\frac{3}{2}}(ry)^{\frac{1}{2}} = 2\sqrt{2ry} \dots \dots (2.)$$

hence (Art. **111**), *The radius of curvature is double the normal;* therefore, when the generating circle moves from A towards M, any radius of curvature, as PP', will be double the normal PN.

If, in Equation (2), we make $y = 0$, we have, $R = 0$;

If we make $y = MB = 2r$, we have, $R = 4r$.

That is, the radius of curvature is zero at the point A, and twice the diameter of the generating circle at B.

Since the radius of curvature and evolute are both zero at the point A, and since they increase equally (Art. **151**), it follows that the length of the evolute $AP'A'$ is equal to $A'MB$, or *twice the diameter of the generating circle.*

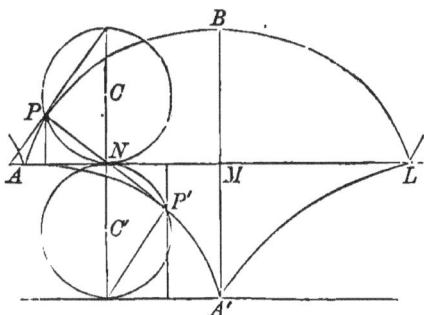

When the point of contact, N, shall have reached M, the point P, will have described the involute APB, and the point P', the evolute $AP'A'$. If we describe a circle on $A'M = \frac{1}{2}A'B$, it will be equal to the generating circle of the cycloid, and the two circles will touch each other at M. Draw $A'X$ parallel to AL.

If now we suppose the circle whose centre is C, to roll along the base from M to A, and the circle whose centre is C', to roll from A' to X, keeping the centres C and C', in a line perpendicular to the base AL, the point P, of the upper circle, will re-describe the involute BPA, and the point P', will re-describe the evolute $A'P'A$. But since the generating circles are equal, and since they are rolled over equal bases, the curves generated will be equal; hence, *the involute and evolute are equal curves.*

The part of the involute beginning at A, is identical with the part of the evolute beginning at A'.

Since the involute and evolute are equal, the length of the involute APB, is equal to twice the diameter of the generating circle; or the length of the entire cycloidal arc $APBL$, is equal to the entire evolute $AP'A'L$, or to four times the diameter of the generating circle.

Equation of the Evolute.

153. The equation of the evolute may be readily found by combining the equations,

$$y - \beta = -\frac{dx^2 + dy^2}{d^2y}, \qquad x - \alpha = \frac{dy(dx^2 + dy^2)}{dx\,d^2y},$$

with the equation of the involute curve.

1st. Find, from the equation of the involute, the values of

$$\frac{dy}{dx} \qquad \text{and} \qquad d^2y,$$

and substitute them in the last two equations; there will result two new equations, involving α, β, x, and y.

2d. Combine these equations with the equation of the involute, and eliminate x and y; the resulting equation will contain α, β, and constants, and will be the equation of the evolute curve.

Evolute of the common Parabola.

154. Let us take, as an example, the common parabola, of which the equation is,

$$y^2 = mx.$$

We shall then have,

$$\frac{dy}{dx} = \frac{m}{2y}, \qquad d^2y = -\frac{m^2 dx^2}{4y^3};$$

and hence,

$$y - \beta = \frac{4y^3}{m^2}\left(\frac{4y^2 + m^2}{4y^2}\right) = \frac{4y^3 + m^2 y}{m^2} = \frac{4y^3}{m^2} + y;$$

and observing that the value of $x - \alpha$ is equal to that of $y - \beta$ multiplied by $-\dfrac{dy}{dx}$, we have,

$$x - \alpha = -\frac{4y^2 + m^2}{2m};$$

hence we have,

$$-\beta = \frac{4y^3}{m^2}, \qquad \text{and} \qquad x - \alpha = -\frac{2y^2}{m} - \frac{m}{2};$$

substituting for y its value in the equation of the involute,

$$y = m^{\frac{1}{2}} x^{\frac{1}{2}},$$

we obtain,

$$-\beta = \frac{4x^{\frac{3}{2}}}{m^{\frac{1}{2}}}; \qquad x - \alpha = -2x - \frac{m}{2};$$

and by eliminating x, we have,

$$\beta^2 = \frac{16}{27m}\left(\alpha - \frac{1}{2}m\right)^3,$$

which is the equation of the evolute.

If we make $\beta = 0$, we have,

$$\alpha = \frac{1}{2}m;$$

and hence, the evolute meets the axis of abscissas at a
distance from the origin equal to half the parameter.
If the origin of co-ordinates be
transferred from A to this point,
we shall have,

$$\alpha' = \alpha - \frac{1}{2}m,$$

and consequently,

$$\beta^2 = \frac{16}{27m}\alpha'^3.$$

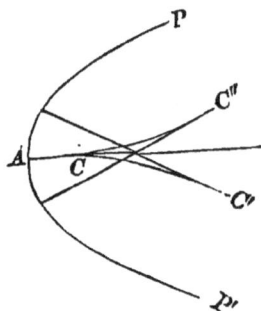

The equation of the curve shows that it is symmetrical
with respect to the axis of abscissas, and that it does not
extend in the direction of the negative values of α'. The
evolute CC' corresponds to the part AP of the involute,
and CC'' to the part AP'. Both are convex towards the
axis of X.

INTEGRAL CALCULUS.

155. In the Differential Calculus, we have developed a system of principles, and given a series of rules, by means of which we deduce, from any given function, two others; the first of which is called the Differential coefficient, and the second, the Differential (Art. **25**). In the Integral Calculus, we have to return from the differential, to the function from which it was derived.

This operation, as a fundamental problem, involves the summation of a series of an infinite number of terms, each of which is infinitely small (Art. **56**). No general rule for the summation of such a series has yet been discovered; and hence, we are obliged to resort, in each particular case, to the operation of reducing the given differential to some equivalent one, whose integral is known (Art. **34**).

Forms of differentials having known Algebraic Functions.

156. We have found (Art. **35**), that every differential monomial of the form,

$$Ax^m dx,$$

in which m is any real number, except -1, may be immediately integrated; and when $m = -1$, the differential becomes that of a logarithmic function, and its integral is $A l x$ (Art. **89**).

157. We have seen that every differential binomial of the form,

$$(a + bx^n)^m x^{n-1} dx,$$

in which the exponent of the variable without the parenthesis is less by 1 than the exponent of the variable within, can be immediately integrated (Art. **41**).

158. We have seen that every function of the form,

$$X dx,$$

in which X can be developed into a series in terms of the ascending powers of x, has an approximate integral which may be readily found (Art. **42**).

Forms of differentials having known Logarithmic Functions.

159. Any function of the form,

$$A \frac{dx}{x},$$

in which the numerator is the differential of the denominator, can be immediately integrated, since the integral is equal to $A l x$ (Art. **89**). In Art. **91**, we have given five other forms of differentials, whose corresponding functions are logarithms.

Forms of differentials having known Circular Functions.

160. In Art. **99**, we have found sixteen differential expressions, each of which has a known integral corresponding to it, and which, being differentiated, will of course produce the given differential.

In all the classes of functions, any differential expression may be considered as integrated, when it is reduced to one of the known forms; and the operations of the Integral Calculus consist, mainly, in making such transformations of given differential expressions, as shall reduce them to equivalent ones, whose integrals are known.

INTEGRATION OF RATIONAL FRACTIONS.

161. Every rational fraction may be written under the form,

$$\frac{P\,x^{n-1} + Q\,x^{n-2}\;\ldots\;+ R\,x + S}{P'x^n\; + Q'x^{n-1}\;\ldots\;+ R'x + S'}\,dx,$$

in which the exponent of the highest power of the variable in the numerator is less by 1 than in the denominator. For, if the greatest exponent in the numerator was equal to, or exceeded the greatest exponent in the denominator, a division might be made, giving one or more entire terms for a quotient, and a remainder, in which the exponent of the leading letter would be less by at least 1, than the exponent of the leading letter in the divisor. The entire terms could then be integrated, and there would remain a fraction under the above form.

1. Let it be required to integrate the expression,

$$\frac{2adx}{x^2 - a^2}.$$

By decomposing the denominator into its factors, we have,

$$\frac{adx}{x^2 - a^2} = \frac{2adx}{(x - a)(x + a)}.$$

Let us make,

$$\frac{2adx}{(x - a)(x + a)} = \left(\frac{A}{x - a} + \frac{B}{x + a}\right)dx,$$

in which A and B are constants, whose values may be found by the method of indeterminate co-efficients.* To find these constants, reduce the terms of the second member of the equation to a common denominator; we shall then have,

$$\frac{adx}{(x - a)(x + a)} = \frac{(Ax + Aa + Bx - Ba)dx}{(x - a)(x + a)}.$$

Comparing the two members of the equation, we find,

$$2a = Ax + Aa + Bx - Ba;$$

or, by arranging with reference to x,

$$(A + B)x + (A - B - 2)a = 0; \text{ hence,}$$

$$A + B = 0, \quad \text{and} \quad (A - B - 2)a = 0;$$

whence, $A = 1$, $B = -1$.

* Bourdon, Art. **194.** University, Art. **180.**

Substituting these values for A and B, we obtain,

$$\frac{2adx}{x^2 - a^2} = \frac{dx}{x - a} - \frac{dx}{x + a};$$

integrating, we find (Art. **89**),

$$\int \frac{adx}{x^2 - a^2} = l(x - a) - l(x + a) + C; \text{ consequently,}$$

$$\int \frac{adx}{x^2 - a^2} = l\left(\frac{x - a}{x + a}\right) + C.$$

2. Find the integral of,

$$\frac{3x - 5}{x^2 - 6x + 8} dx.$$

Resolving the denominator into its two binomial factors, $(x - 2)$, and $(x - 4)$, we have,

$$\frac{3x - 5}{x^2 - 6x + 8} = \frac{A}{x - 2} + \frac{B}{x - 4}; \text{ hence,}$$

$$\frac{3x - 5}{x^2 - 6x + 8} = \frac{Ax - 4A + Bx - 2B}{x^2 - 6x + 8};$$

by comparing the coefficients of x, we have,

$$- 5 = - 4A - 2B, \qquad 3 = A + B,$$

which gives, $\qquad B = \frac{7}{2}, \qquad A = -\frac{1}{2};$

substituting these values, we have,

$$\int \frac{3x - 5}{x^2 - 6x + 8} dx = -\frac{1}{2} \int \frac{dx}{x - 2} + \frac{7}{2} \int \frac{dx}{x - 4} + C$$

$$= \frac{7}{2} \log(x - 4) - \frac{1}{2} \log(x - 2) + C.$$

Hence, for the integration of rational fractions:

1st. *Resolve the fraction into partial fractions, of which the numerators shall be constants, and the denominators factors of the denominator of the given fraction.*

2d. *Find the values of the numerators of the partial fractions, and multiply each by dx.*

3d. *Integrate each partial fraction separately, and the sum of the integrals thus found will be the integral sought.*

INTEGRATION BY PARTS.

162. The integration of differentials is often effected by resolving them into two parts, of which one has a known integral.

We have seen (Art. **27**), that,

$$d(uv) = udv + vdu,$$

whence, by integrating,

$$uv = \int udv + \int vdu,$$

and, consequently,

$$\int udv = uv - \int vdu.$$

Hence, if we have a differential of the form Xdx, which can be decomposed into two factors P and Qdx, of which one of them, Qdx, can be integrated, we shall have, by making $\int Qdx = v$, and $P = u$,

$$\int Xdx = \int PQdx = \int udv = uv - \int vdu \quad . \quad (1.)$$

in which it is only required to integrate the term $\int vdu$.

EXAMPLES.

1. Integrate the expression, $x^3 dx \sqrt{a^2 + x^2}$.

This may be divided into the two factors,

$$x^2, \quad \text{and} \quad x dx \sqrt{a^2 + x^2},$$

of which the second is integrable (Art. **41**).

Put, $\quad u = x^2, \quad$ and $\quad dv = x dx \sqrt{a^2 + x^2};$ then,

$$du = 2x dx, \quad \text{and} \quad v = \int x dx \sqrt{a^2 + x^2} = \frac{(a^2 + x^2)^{\frac{3}{2}}}{3}.$$

Substituting these values in Formula (1),

$$\int u dv = x^2 \left(\frac{a^2 + x^2}{3}\right)^{\frac{3}{2}} - \int \frac{(a^2 + x^2)^{\frac{3}{2}}}{3} \times 2x dx;$$

and finally,

$$\int x^3 dx \sqrt{a^2 + x^2} = x^2 \left(\frac{a^2 + x^2}{3}\right)^{\frac{3}{2}} - \frac{2}{15}(a^2 + x^2)^{\frac{5}{2}} + C.$$

2. Integrate the expression, $\dfrac{x^2 dx}{(a^2 - x^2)^{\frac{3}{2}}}$.

The two factors are, $\quad x, \quad$ and $\quad x dx (a^2 - x^2)^{-\frac{3}{2}}$.

$$u = x; \quad dv = x dx (a^2 - x^2)^{-\frac{3}{2}}; \quad v = \frac{1}{\sqrt{(a^2 - x^2)}} \cdot$$

$$\int u dv = \frac{u}{\sqrt{a^2 - x^2}} + \int \frac{dx}{\sqrt{a^2 - x^2}}; \quad \text{whence,}$$

$$\int \frac{x^2 dx}{(a^2 - x^2)^{\frac{3}{2}}} = \frac{x}{\sqrt{a^2 - x^2}} + \sin^{-1} \frac{x}{a} \quad \text{(Art. **99**).}$$

INTEGRATION OF BINOMIAL DIFFERENTIALS.

Form of Binomial.

163. Every binomial differential may be placed under the form,

$$x^{m-1}dx(a + bx^n)^p,$$

in which m and n are whole numbers, and n positive; and in which p is entire or fractional, positive or negative.

1. For, if m and n are fractional, the binomial takes the form,

$$x^{\frac{1}{3}}dx(a + bx^{\frac{1}{2}})^p.$$

If we make $x = z^6$, that is, if we substitute for x, another variable, z, with an exponent equal to the least common multiple of the denominators of the exponents of x, we shall have,

$$x^{\frac{1}{3}}dx(a + bx^{\frac{1}{2}})^p = 6z^7dz(a + bz^3)^p,$$

in which the exponents of the variable are entire.

2. If n is negative, we have,

$$x^{m-1}dx(a + bx^{-n})^p,$$

and by making $x = \frac{1}{z}$, we obtain,

$$- z^{-m+1}dz(a + bz^n)^p,$$

in which n is positive.

3. If x enters into both terms of the binomial, giving the form,

$$x^{m-}dx(ax^r + bx^n)^p,$$

in which the lowest power of x is written in the first term, we divide the binomial within the parenthesis by x^r, and multiply the factor without by x^{rp}; this gives,

$$x^{m+pr-1} dx (a + bx^{n-r})^p,$$

which is of the required form when the exponent $m + p^{r-1}$, is a whole number, and may easily be reduced to it, when that exponent is fractional.

When a Binomial can be integrated.

164.—1. If p is entire and positive, it is plain that the binomial can be integrated. For, when the binomial is raised to the indicated power, there will be a finite number of terms, each of which, after being multiplied by $x^{m-1} dx$, may be integrated (Art. **35**).

2. If $m = n$, the binomial can be integrated (Art. **41**)

3. If p is entire, and negative, the binomial will take the form,

$$\frac{x^{m-1} dx}{(a + bx^n)^p};$$

which is a rational fraction.

Formula A.

For diminishing the exponent of the variable without the parenthesis.

165. Let us resume the differential binomial,

$$x^{m-1} dx (a + bx^n)^s$$

If we multiply by the two factors, x^n and x^{-n}, the value will not be changed, and we obtain,

$$x^{m-n}x^{n-1}dx(a + bx^n)^p.$$

Now, the factor $x^{n-1}dx(a + bx^n)^p$ is integrable, whatever be the value of p (Art. **41**). Denoting the first factor, x^{m-n} by u, and the second by dv, we have,

$$du = (m - n)x^{m-n-1}dx, \quad \text{and} \quad v = \frac{(a + bx^n)^{p+1}}{(p + 1)nb};$$

and, consequently,
$$\int x^{m-1}dx(a + bx^n)^p =$$

$$\frac{x^{m-n}(a + bx^n)^{p+1}}{(p + 1)nb} - \frac{m - n}{(p + 1)nb}\int x^{m-n-1}dx(a + bx^n)^{p+1}.$$

But,
$$\int x^{m-n-1} dx(a + bx^n)^{p+1} =$$

$$\int x^{m-n-1} dx(a + bx^n)^p (a + bx^n) =$$

$$a\int x^{m-n-1} dx(a + bx^n)^p + b\int x^{m-1} dx(a + bx^n)^p;$$

substituting this last value in the preceding equation, and collecting the terms containing,

$$\int x^{m-1} dx(a + bx^n)^p,$$

we have,
$$\left(1 + \frac{m - n}{(p + 1)n}\right)\int x^{m-1} dx(a + bx^n)^p =$$

$$\frac{x^{m-n}(a + bx^n)^{p+1} - a(m - n)\int x^{m-n-1} dx(a + bx^n)^p}{(p + 1)nb};$$

whence,

$$(\mathbf{A}) \ldots \ldots \ldots \ldots \int x^{m-1} dx (a + bx^n)^p =$$

$$\frac{x^{m-n}(a + bx^n)^{p+1} - a(m - n)\int x^{m-n-1} dx (a + bx^n)^p}{b(pn + m)}.$$

This formula reduces the differential binomial,

$$\int x^{m-1} dx (a + bx^n)^p, \qquad \text{to} \qquad \int x^{m-n-1} dx (a + bx^n)^p;$$

and by a similar operation, we should find,

$$\int x^{m-n-1} dx (a + bx^n)^p, \text{ to depend on, } \int x^{m-2n-1} dx (a + bx^n)^q;$$

consequently, *each operation diminishes the exponent of the variable without the parenthesis by the exponent of the variable within.*

After the second integration, the factor $m - n$, of the second term, becomes $m - 2n$; and after the third, $m - 3n$, &c. If m is a multiple of n, the factor $m - n$, $m - 2n$, $m - 3n$, &c., will finally become equal to 0, and then the differential into which it is multiplied will disappear, and the given differential can be integrated. Hence, *a differential binomial can be integrated, when the exponent of the variable without the parenthesis plus 1, is a multiple of the exponent within.*

<h3 style="text-align:center">APPLICATIONS.</h3>

166. We have frequent occasion to integrate differential binomials of the form,

$$\frac{x^m dx}{\sqrt{a^2 - x^2}} = x^m dx (a^2 - x^2)^{-\frac{1}{2}}.$$

The differential binomial $x^{m-1} dx (a + bx^n)^p$ will assume this form, if we substitute,

for m,   .   .   . $m + 1$;
" a,   .   .   . a^2;
" b^2, .   .   . -1;
" n,   .   .   . 2;
" p,   .   .   . $-\frac{1}{2}$.

Making these substitutions in Formula A, we have,

$$\int \frac{x^m\, dx}{\sqrt{a^2 - x^2}} = - \frac{x^{m-1}}{m} \sqrt{a^2 - x^2} + \frac{a^2 (m - 1)}{m} \int \frac{x^{m-2}\, dx}{\sqrt{a^2 - x^2}};$$

so that the given binomial differential depends on,

$$\int \frac{x^{m-2}\, dx}{\sqrt{a^2 - x^2}};$$

and in a similar manner this is found to depend upon,

$$\int \frac{x^{m-4}\, dx}{\sqrt{a^2 - x^2}};$$

and so on, each operation diminishing the exponent of x by 2. If m is an even number, the integral will depend, afte $\frac{m}{2}$ operations, on that of,

$$\int \frac{dx}{\sqrt{a^2 - x^2}} = \sin^{-1} \frac{x}{a} \quad \text{(Art. } 99\text{)}.$$

FORMULA 𝔅.

For diminishing the exponent of the parenthesis.

167. By changing the form of the given differential binomial, we have,

$$\int x^{m-1}\,dx(a+bx^n)^p =$$

$$\int x^{m-1}\,dx(a+bx^n)^{p-1}(a+bx^n) =$$

$$a\int x^{m-1}\,dx(a+bx^n)^{p-1} + b\int x^{m+n-1}\,dx(a+bx^n)^{p-1}.$$

Applying Formula 𝔄 to the second term, and observing that m is changed to $m+n$, and p to $p-1$, we have,

$$\int x^{m+n-1}\,dx(a+bx^n)^{p-1} =$$

$$\frac{x^m(a+bx^n)^p - am\int x^{m-1}\,dx(a+bx^n)^{p-1}}{b(pn+m)}.$$

Substituting this value in the last equation, we have,

$$(\mathfrak{B})\ldots\ldots\ldots\ldots\int x^{m-1}\,dx(a+bx^n)^p =$$

$$\frac{x^m(a+bx^n)^p + pna\int x^{m-1}\,dx(a+bx^n)^{p-1}}{pn+m},$$

in which the exponent of the parenthesis is diminished by 1, for each operation.

APPLICATIONS.

1. Integrate the expression $dx(a^2+x^2)^{\frac{3}{2}}$.

The differential binomial $x^{m-1}\,dx(a+bx^n)^p$ will assume this form, if we make $m = 1$, $a = a^2$, $b = 1$, $n = 2$, and $p = \frac{3}{2}$.

Substituting these values in the formula, we have,

$$\int dx(a^2 + x^2)^{\frac{3}{2}} = \frac{x(a^2 + x^2)^{\frac{3}{2}} + 3a^2\int dx(a^2 + x^2)^{\frac{1}{2}}}{4}.$$

Applying the formula a second time, we have,

$$\int dx(a^2 + x^2)^{\frac{1}{2}} = \frac{x(a^2 + x^2)^{\frac{1}{2}}}{2} + \frac{a^2}{2}\int \frac{dx}{\sqrt{a^2 + x^2}}.$$

But we have found (Art. **91**),

$$\int \frac{dx}{\sqrt{a^2 + x^2}} = l\left(x + \sqrt{a^2 + x^2}\right);$$

hence, $$\int dx(a^2 + x^2)^{\frac{3}{2}} =$$

$$\frac{x(a^2 + x^2)^{\frac{3}{2}}}{4} + 3a^2 x\frac{(a^2 + x^2)^{\frac{1}{2}}}{8} + \frac{3a^4}{8}\cdot l\left(x + \sqrt{a^2 + x^2}\right) + C.$$

2. Integrate the expression, $dx\sqrt{r^2 - x^2}$.

The first member of the equation will assume this form, if we make, $m = 1$, $a = r^2$, $b = -1$, $n = 2$, and $p = \frac{1}{2}$. Substituting these values in the formula, we have,

$$\int dx\sqrt{r^2 - x^2} = \frac{1}{2}x(r^2 - x^2)^{\frac{1}{2}} + \frac{1}{2}r^2\int \frac{dx}{\sqrt{r^2 - x^2}};$$

whence, by substitution (Art. **99**),

$$\int dx\sqrt{r^2 - x^2} = \frac{1}{2}x(r^2 - x^2)^{\frac{1}{2}} + \frac{1}{2}r^2 \sin^{-1}\frac{x}{r} + C.$$

Formula C.

For diminishing the exponent of the variable without the parenthesis, when it is negative.

168. It is evident that Formula A will only diminish $m - 1$, the exponent of the variable, when m is positive. We are now to determine a formula for diminishing this exponent when m is negative.

From Formula A, we deduce,

$$\int x^{m-n-1} dx (a + bx^n)^p =$$

$$\frac{x^{m-n}(a + bx^n)^{p+1} - b(m + np) \int x^{m-1} dx (a + bx^n)^p}{a(m - n)};$$

changing m, to $-m + n$, we have,

$$(C) \cdot \cdot \cdot \cdot \cdot \cdot \cdot \cdot \cdot \int x^{-m-1} dx (a + bx^n)^p =$$

$$\frac{x^{-m}(a + bx^n)^{p+1} + b(m - n - np) \int x^{-m+n-1} dx (a + bx^n)^p}{- am},$$

in which formula, it should be remembered that the negative sign has been attributed to the exponent m.

APPLICATIONS.

1. Integrate the expression $x^{-2} dx (2 - x^2)^{-\frac{3}{2}}$.

The first member of Equation (C) will assume this form, if we make $m = 1$, $a = 2$, $b = -1$, $n = 2$, and $p = -\frac{3}{2}$. Substituting these values, we have,

$$\int x^{-2} dx (2 - x^2)^{-\frac{3}{2}} = - \frac{x^{-1}(2 - x^2)^{-\frac{1}{2}}}{2} + \int (2 - x^2)^{-\frac{3}{2}} dx.$$

The differential term in the second member will be integrated by the next formula.

FORMULA 𝔇.

For diminishing the exponent of the parenthesis when it is negative.

169. It is evident that Formula 𝔅 will only diminish p, the exponent of the parenthesis, when p is positive. We are now to determine a formula for diminishing this exponent when p is negative.

We find, from Formula 𝔅,

$$\int x^{m-1} dx (a + bx^n)^{p-1} =$$

$$\frac{- x^m (a + bx^n)^p + (m + np) \int x^{m-1} dx (a + bx^n)^p}{pna};$$

writing for p, $-p + 1$, we have,

$$(\mathfrak{D}) \quad . \quad . \quad . \quad . \quad . \quad . \quad . \quad . \quad \int x^{m-1} dx (a + bx^n)^{-p} =$$

$$\frac{x^m (a + bx^n)^{-p+1} - (m + n - np) \int x^{m-1} dx (a + bx^n)^{-p+1}}{na(p-1)} .$$

When $p = 1$, $p - 1 = 0$; the second member becomes infinite, and the given expression becomes a rational fraction.

APPLICATIONS.

1. Integrate the expression, $\int dx (2 - x^2)^{-\frac{3}{4}}$.

The first member of Equation ⑩ will assume this form, if we make $m = 1$, $a = 2$, $b = -1$, $n = 2$, and $p = -\frac{3}{2}$. Substituting these values, we have,

$$\int dx(2 - x^2)^{-\frac{3}{2}} = \frac{x(2 - x^2)^{-\frac{1}{2}}}{2};$$

since the coefficient of the second term, in the formula, becomes zero.

Returning, then, to the example under the last formula, we have,

$$\int x^{-2}dx(2 - x^2)^{-\frac{3}{2}} = -\frac{x^{-1}(2 - x^2)^{-\frac{1}{2}}}{2} + \frac{x(2 - x^2)^{-\frac{1}{2}}}{2} + C.$$

2. By means of Formula ⑩, we are able to integrate the expression,

$$\frac{dz}{(a^2 + z^2)^p} = dz(a^2 + z^2)^{-p},$$

when p is a whole number.

The general formula will assume this form, if we make $m = 1$, $x = z$, $a = a^2$, $b = 1$, $n = 2$.

Each application of the formula will reduce the exponent $- p$, by 1, until the integral will finally depend on that of

$$\frac{dz}{a^2 + z^2} = \frac{1}{a} \tan^{-1}\frac{z}{a} + C \quad \text{(Art. 99)}.$$

FORMULA ⑫.

When the variable enters into both terms of the binomial.

170. Let it be required to integrate the expression,

$$\frac{x^q dx}{\sqrt{2ax - x^2}} = x^q dx(2ax - x^2)^{-\frac{1}{2}}.$$

The second member may be placed under the form,

$$\int x^{q-\frac{1}{2}}dx(2a - x)^{-\frac{1}{2}}.$$

We apply Formula $\mathbb{A}$, by making,

$$m = q + \frac{1}{2}, \qquad n = 1, \qquad p = -\frac{1}{2}, \qquad a = 2, \qquad b = -1;$$

we shall then have,

$$\int x^{q-\frac{1}{2}}dx(2a - x)^{-\frac{1}{2}} =$$

$$-\frac{x^{q-\frac{1}{2}}(2a - x)^{\frac{1}{2}}}{q} + \frac{2a(q - \frac{1}{2})}{q}\int x^{q-\frac{3}{2}}dx(2a - x)^{-\frac{1}{2}}.$$

If we observe that,

$$x^{q-\frac{1}{2}} = x^{q-1}x^{\frac{1}{2}}, \qquad \text{and} \qquad x^{q-\frac{3}{2}} = x^{q-1}x^{-\frac{1}{2}},$$

and pass the fractional powers of x within the parentheses, we shall have,

$$(\mathbb{B}) \quad . \quad . \quad . \quad . \quad . \quad . \quad . \quad . \quad . \quad . \quad \int \frac{x^q dx}{\sqrt{2ax - x^2}} =$$

$$-\frac{x^{q-1}\sqrt{2ax - x^2}}{q} + \frac{(2q - 1)a}{q}\int \frac{x^{q-1}dx}{\sqrt{2ax - x^2}}.$$

Each application of this formula diminishes the expo-nent of the variable without the parenthesis by 1. If q is a positive and entire number, we shall have, after q reductions,

$$\int \frac{dx}{\sqrt{2ax - x^2}} = \text{ver-sin}^{-1}\frac{x}{a} + C \quad \text{(Art. \textbf{99})}.$$

www.ingramcontent.com/pod-product-compliance
Lightning Source LLC
Chambersburg PA
CBHW021701210326
41599CB00013B/1478